Henry Jellett

A Short Practice of Midwifery

Embodying the Treatment Adopted in the Rotunda Hospital, Dublin

Henry Jellett

A Short Practice of Midwifery
Embodying the Treatment Adopted in the Rotunda Hospital, Dublin

ISBN/EAN: 9783337163211

Printed in Europe, USA, Canada, Australia, Japan

Cover: Foto ©berggeist007 / pixelio.de

More available books at **www.hansebooks.com**

A SHORT PRACTICE OF MIDWIFERY.

A SHORT PRACTICE OF MIDWIFERY,

EMBODYING THE TREATMENT ADOPTED IN
THE ROTUNDA HOSPITAL, DUBLIN.

BY

HENRY JELLETT, B.A., M.D., B.Ch., B.A.O. (Dublin University), L.R.C.P.I., L.M.;

ASSISTANT MASTER, ROTUNDA HOSPITAL.

WITH A PREFACE BY

W. J. SMYLY, M.D., F.R.C.P.I.,

LATE MASTER OF THE ROTUNDA HOSPITAL.

*With 45 Illustrations, and an Appendix containing the statistics
of the Hospital for the last seven years.*

PHILADELPHIA
P. BLAKISTON, SON & CO.
1012, WALNUT STREET
1897

PREFACE.

I AM sure that Dr Jellett's little book will prove
acceptable to many practitioners and students who
desire a succinct account of the methods adopted in
the Rotunda Hospital, in the management of parturient women. In many particulars the views expressed are at variance with the rules laid down in
most text-books, and I may here emphasize a few of
these. It has been shown, that, whereas in hospitals
the introduction of antiseptics has been followed by
most gratifying results, in private practice little if
any improvement is observable. To account for this
deplorable state of affairs, it has been pointed out that
proper precautions are not so universally adopted by
practitioners and nurses as they should be; and, also,
on the other hand, that too much reliance upon
antiseptic methods has encouraged " meddlesome
midwifery; " so that what has been gained by the
former has been sacrificed by the latter. The
recommendations in this work regarding the substi-

tution, as far as possible, of external for internal manipulations; the avoidance of routine douching, of the use of the plug in abortions and placenta prævia, and of the forceps in cases where the head has not passed the pelvic brim; and the management of the third stage of labour, are matters of the greatest importance. I am entirely in accord with the statement that a practitioner can appraise his own merits by the infrequency of post-partum hæmorrhage in his practice.

The subjects, which will probably provoke most criticism, are the methods of treatment recommended in accidental hæmorrhage and eclampsia. In the first two years of my Mastership, I treated all serious cases of accidental hæmorrhage by rupturing the membranes; and, if that did not prove effectual, delivery was effected by version and extraction or perforation. The results were so bad that I resorted to plugging in all cases of external accidental hæmorrhage in which the membranes were intact, and labour pains absent or feeble—that is in the great majority of cases—and with excellent results. The fear that an external would be converted into an internal hæmorrhage proved groundless. The use of chloroform in puerperal eclampsia I abandoned with the greatest reluctance. Nothing is more gratifying to the practitioner himself and the relatives of the patient than the complete control of the convulsions by chloroform, but it does not save the patient's life ; on the contrary, it increases the tendency to death.

Even to those who differ from the views advanced, this little work will afford matter for reflection, especially as the results of the treatment advocated can be judged from the statistics appended.

W. J. SMYLY, M.D., F.R.C.P.I.,

Ex-Master of the Rotunda Hospital. Dublin.

AUTHOR'S PREFACE

THE following short Practice of Midwifery has been written with the object of giving a concise and practical description of the treatment, which has been so successfully carried out in this Hospital, under the mastership of Dr W. J. Smyly; and which is still followed by his successor, Dr R. D. Purefoy. I feel that it is necessary to offer an excuse, for the publication of another handbook of midwifery, when so very many are in existence. My excuse is twofold. I have written this little work, first, because I was requested to do so by several of the students attending the Hospital; secondly, because the extremely successful results, following the treatment adopted here, seem to warrant its publication. I have given, in the form of an appendix, statistics of the various cases treated in the Hospital, during the mastership of Dr W. J. Smyly. It will be seen, from them, what a striking improvement has occurred in the rate of maternal mortality. Necessarily, in the space of seven years, considerable changes have been made in the various methods of treatment. The treatment, which I describe, is that which has been selected in accordance with the experience of these years; and which, during the last year, has yielded the extremely low maternal mortality of 0·06 per cent.

I should like to say a few words in apology for the shortcomings of the chapter on Anatomy. When I commenced this book, I determined to omit any such chapter, as there are so many excellent descriptions of pelvic anatomy in the various text-books of midwifery. As far as students and practitioners are concerned, I have adhered to that determination. The short chapter on pelvic anatomy is not intended for them. It is solely meant for the nurses of this Hospital, who may read the book for their examinations; and to whom, I believe, such a chapter will be useful.

I wish here to express my deep sense of the kindness of Dr Smyly, in permitting me to publish his teaching in this form. I trust, that the extreme clearness and science of his teaching will not be so obscured by my shortcomings, as to render it less obvious to those who read it in this book, than it has been to those who have heard it from himself. Further, I wish to express my thanks to the several gentlemen who have assisted me; to Dr Carton, who has devoted so much time and trouble to the illustrations; to Dr Drury and Mr Bresland, who have corrected the proof sheets; and to Dr Purefoy, who has permitted me to associate the book with the Rotunda Hospital. Lastly, I must express my indebtedness to the works of Winckel, Lusk, and Dührssen, from all of which I have derived great assistance. Many of the illustrations are new, and such are the work of Dr Carton. The remainder are either directly copied from other works, or slightly modified in the copying. All such illustrations are acknowledged in the text beneath them; for the majority of them, I am indebted to the "Norris Text-book of Obstetrics."

HENRY JELLETT.

ROTUNDA HOSPITAL, DUBLIN;
March, 1897.

CONTENTS.

CHAPTER IX.

NATURAL PRESENTATIONS.

CHAPTER X.

UNNATURAL PRESENTATIONS.

CHAPTER XI.

MULTIPLE PREGNANCY.

CHAPTER XII.

THE PUERPERIUM.

CHAPTER XXVI.

CONTRACTED PELVIS.

CHAPTER XXVII.

PROLAPSE OF THE CORD.

CHAPTER XXVIII.

ECLAMPSIA.

CHAPTER XXIX.

SEPTIC INFECTION.

A SHORT PRACTICE OF MIDWIFERY.

CHAPTER I.

ASEPSIS IN MIDWIFERY.

Importance of Asepsis in Midwifery—Mechanism by which the Uterus is kept Aseptic during Pregnancy and Labour—The Vaginal Secretion—The Operculum—Method of sterilising the Hands and the Instruments — Prophylactic Douching — Douching Solutions—Preparation of Patient for an Obstetrical Operation.

It is not an exaggeration to say that the most essential knowledge in midwifery is the knowledge of asepsis. A practitioner who knows nothing of the science and art of midwifery, except that it is absolutely necessary that his hands and instruments be sterile, will save more lives than the most accomplished obstetrician who does not practise asepsis. It is therefore most fitting that the first chapter of this little book should deal fully with the science of asepsis in midwifery.

If there were no such things as vaginal examinations, or as intra-vaginal or intra-uterine operations, a previously healthy patient would never suffer from acute sepsis. This being so, there must be some natural mechanism, which prevents the entrance of pyogenic organisms into the uterus. At the com-

1

mencement of labour, a healthy vagina is lubricated with a fluid, which is composed partly of the secretion of the cervical glands, and partly of serous transudation from the vaginal blood-vessels. This fluid is swarming with bacilli, which not only are not pathological, but are a direct bar to the entrance of pathological bacteria. This they effect by the generation of lactic acid, which renders the vaginal discharge acid, and so prevents the development of pyogenic organisms, as the latter can only exist in an alkaline medium. It has been found by experiment, that pyogenic organisms introduced into the vagina are destroyed in a few hours. In addition to the protection furnished by the vaginal discharge, there is a still further bar to the entrance of bacteria into the uterus; this is the plug of mucus which fills the cervix, the so-called operculum. This plug is described as consisting of three layers,—an upper or uterine layer, a middle or cervical layer, and a lower or vaginal layer. The upper layer contains no bacteria of any kind, hence it is aseptic. The middle layer contains dead bacteria and quantities of white corpuscles. These latter act as phagocytes, and hence the middle layer is antiseptic. The lower layer contains swarms of bacteria,—non-pathological if the vagina be healthy, and pathological if there be any form of vaginitis present. It therefore is septic. It is said, that no bacteria can find their way past the middle layer of the operculum, except gonococci. Thus, by the aid of the vaginal bacilli and of the operculum, the uterus is kept aseptic before delivery. After the birth of the child all bacteria have disappeared from the vagina. This is brought about in the following manner :—When the membranes rupture, the flow of liquor amnii through the vagina, washes the greater quantity of micro-organisms out of it. Then the presenting part of the child, as it passes through the vagina, distends its walls to the utmost, so that the second rush of liquor amnii is enabled to wash away all that remain. Thus the uterus is prevented from becoming infected after delivery, the time at which it is exposed to the greatest risks.

Inasmuch as vaginal examinations and operations must be performed, it is incumbent on us to do everything in our power to avoid the introduction of germs. They may be introduced in three ways :— (1) by septic hands, (2) by septic instruments, (3) by carrying up septic matter from the vulva or vagina on our fingers or instruments into the uterus.

To avoid the first the hands must be cleansed thoroughly. The following method has stood the test of time in the Rotunda Hospital, and appears to be sufficient :—First scrub the hands thoroughly in soap and water with a strong nail-brush, paying particular attention to the nails. Then rinse the hands in plain water to remove all trace of soap, which decomposes corrosive sublimate. Lastly, immerse the hands for one minute in a 1 in 500 solution of corrosive sublimate. Avoid the use of lubricants if possible. If one must be used, let it be thoroughly aseptic. Carbolised vaseline is never safe, particularly when kept in a box into which dirty fingers are introduced from time to time. Soap, which has been boiled in the making, furnishes an excellent and safe lubricant. It requires one precaution, viz. that the outer layer be first washed off, and thus any dirt

which was in contact with it removed.   The inner
layer is perfectly aseptic.   To avoid the second
method of infection let your instruments be, as far as
possible, of metal, in order that they can be boiled.
This should be done for at least five minutes.   If a
one per cent. solution of common washing soda be
used, the instruments will not become rusty.   To
avoid the third method of introducing infection, the
vulva should be thoroughly washed and disinfected
by the nurse, before any examination is made.   If
any operation has to be performed, which necessitates
the introduction of fingers or of instruments into the
uterus, the vagina must be disinfected as well.   This
is done, because, in many cases, the vaginal discharge is not normal; and also because bacteria, which
may be non-pathogenic in the vagina, may become pathogenic by feeding upon dead tissue,
blood-clots, &c., in the uterus.   To disinfect the
vagina it must be thoroughly douched with a solution of creolin, scrubbed all round with the fingers
and a small piece of soap, and then douched again.

Now as regards prophylactic routine douching,
i. e. douching before and after labour with a view to
preventing sepsis.   Ante-partum douching is quite
unnecessary in a normal case.   It is indicated under
certain conditions :—

(1) If any operation be about to be performed.

(2) If there be any purulent or putrid discharge
from the vagina or the uterus.

(3) If the patient be a very long time in the
second stage.   In these cases, the liquor amnii drains
away slowly, and, by the time the head is born, there
is not sufficient left to wash out the vagina.   Also,

during the protracted labour, some of the liquor amnii may lie in the vagina and decompose.

Prophylactic post-partum douching is a practice which must be strongly condemned. Inasmuch as it has been proved that the vagina is sterile after delivery, it is quite unnecessary. In addition, when it is administered as a routine by an ignorant nurse, with a much-used Higginson's syringe, it is extremely dangerous. Post-partum douching is indicated under certain conditions. It must then be regarded as a serious operation, and performed with the strictest attention to asepsis. If possible it should be administered by the medical attendant himself. The indications for its use are as follow :—

(1) If the hands have been introduced into the uterus, e. g. for the removal of a placenta.

(2) If there be post-partum hæmorrhage.

(3) If the fœtus or placenta be putrid.

(4) If there be purulent discharge from the uterus.

(5) If the lochia become putrid at any time during the puerperium.

For the purpose of douching, I recommend a mixture of creolin and water, of a strength of half an ounce of the former to a gallon of the latter. It should be used at a temperature of 100° F. in ordinary cases, but to check hæmorrhage it may be used up to 120° F. Corrosive sublimate is almost useless for the purpose of douching. Before labour it corrugates the tissues and makes them rigid, so that lacerations are very liable to occur. After delivery, if used too strong, or if any be left behind, it may cause symptoms of mercurial poisoning. While the lochia is red a corrosive douche should never be used stronger than 1 in 7000 for intra-uterine douching. Corrosive sublimate is here useless as an antiseptic, as it decomposes quickly in the presence of albumen. Douches do not remove bacteria from the vagina and uterus by destroying them by means of the antiseptic in the douche, as the fluid does not remain in contact with them for a sufficient length of time. Bacteria are removed mechanically by the flow of fluid, whilst the antiseptic merely helps to render the water in the douche aseptic. Coal-tar derivatives, as creolin and carbolic acid, are said in addition to cause a leucocytosis, so increasing phagocytosis.

We should, therefore, commence every obstetrical operation, in which either hands or instruments have to be introduced into the uterus, in the following manner :—

(1) Wash the external genitals and the skin round them thoroughly with soap and water.

(2) Douche out the vagina thoroughly with creolin solution, then scrub its walls with the fingers and a clean piece of soap, and then douche it out again.

CHAPTER II.

ELEMENTARY PELVIC ANATOMY.

The Bony Pelvis —The Diameters of the Pelvis—The Inclined Planes — The Female Organs of Generation : the Ovaries, Fallopian Tubes, Uterus, Vagina—The Ovum—Table for estimating the Age of the Fœtus by its Length—The Fœtal Skull.

THE bony pelvis is formed by four bones—the two innominate bones, the sacrum, and the coccyx. These articulate in the following manner :—Each innominate bone articulates with the sacrum at the sacro-iliac synchondrosis, and with its fellow at the pubis. The sacrum articulates with the last lumbar vertebra, with the two innominate bones at the sacro-iliac joint, and with the coccyx at the sacro-coccygeal joint. The coccyx articulates with the sacrum alone. The joints are usually rigid, but towards the end of pregnancy their ligaments soften, and so permit slight movements to take place. The sacrum moves in an antero-posterior plane, as if it were pivoted upon the sacro-iliac synchrondroses. As the fœtal head descends, it presses upon the promontory of the sacrum and forces it slightly backwards. As soon as the head has passed the brim the promontory returns to its original position, and then moves slightly forwards, as the descended head drives the lower pieces of the sacrum backwards. The coccyx

can also move backwards on the sacro-coccygeal joint,
and thus increases the antero-posterior diameter of
the outlet by about three-quarters of an inch.   The
pubic bones can separate slightly at the symphysis.

The true pelvis, with which alone we are concerned,
possesses certain diameters which are of great importance.   These are, the diameters of the brim and
the diameters of the outlet.   The brim has four
chief diameters, and by their measurement we can
ascertain its shape and size.   They are :—

(1) *The conjugata vera, i. e.* the distance between
the promontory of the sacrum and the most prominent part of the inner surface of the symphysis
pubis.   It measures normally from 4 to 4¼ inches.

(2) *The transverse diameter, i. e.* the greatest distance between the lateral margins of the brim.   It
measures 5¼ inches.

(3 and 4) *Two oblique diameters*, right and left, *i. e.*
the distance between either sacro-iliac synchondrosis
and the pectineal eminence of the opposite side.
The right oblique diameter runs from the right
sacro-iliac synchondrosis to the left pectineal eminence ; the left oblique diameter runs from the left
sacro-iliac synchondrosis to the right pectineal eminence.   They each measure about 5 inches.

The outlet has two chief diameters :—

(1) *The antero-posterior diameter, i. e.* the distance
from the tip of the coccyx to the lower border of
the symphysis.   It measures 3½ inches, and can be
increased by three-quarters of an inch by the back-
ward movement of the coccyx.

(2) *The transverse, i. e.* the distance between the
tuberosities of the ischii.   It measures 4⅜ inches.

It is also well to know the measurement of the pelvic cavity, first, in its plane of greatest expansion, and, secondly, in its plane of greatest contraction. *The plane of greatest expansion* passes through the middle of the symphysis and the junction of the second and third pieces of the sacrum. *The plane of greatest contraction* passes through the lower margin of the symphysis and the lower margin of the last piece of the sacrum. For the sake of convenience I append a table of the diameters of the pelvis in these various planes :—

	Antero-posterior Diameter.	Trans-verse.	Oblique.
Plane of pelvic inlet .	4—4¼ inches	5¼ inches	5 inches
„ greatest expansion	5$\frac{1}{10}$,,	5 ,,	
„ greatest contraction	4$\frac{3}{5}$,,	4$\frac{1}{5}$,,	
„ pelvic outlet .	3$\frac{1}{5}$,,	4$\frac{2}{5}$,,	

The inclined planes of the pelvis, which concern the obstetrician, are two in number. They start at either side in front of the ischiatic spines, and slope downwards and forwards over the ischium. They are so placed that the part of the child, which first impinges on either of them, is directed downwards and forwards. They are an important factor in the causation of internal rotation.

There are other measurements of the pelvis which are of interest in some cases. I therefore append them :—

The distance between the anterior superior spines = 10 inches.
     „   „   iliac crests   = 11$\frac{2}{5}$,,
The height of the posterior wall of the pelvis   = 5$\frac{1}{5}$,,
     „   anterior   „   „   = 1$\frac{3}{5}$,,
The external conjugate, *i. e.* the distance between the last lumbar spine and the middle of the external surface of the pubes .   .   . } = 8 ,.

The Female Organs of Generation.—The female organs of generation are as follow :—

1. *The Ovaries.*—The ovaries lie, one at each side of the uterus, attached to the posterior aspect of the broad ligament. They measure $1\frac{1}{2}$ inches in length, $\frac{4}{5}$ inch in width, and about $\frac{1}{5}$ inch in thickness. They are uncovered by peritoneum. Their blood-supply is received through the ovarian artery, a branch of the abdominal aorta, and is returned through the ovarian veins. These latter ramify in the broad ligament to form the pampiniform plexus, and then flow, on the right side, into the inferior vena cava; on the left side, into the renal vein. The nerve-supply of the ovaries is derived from the 3rd and 4th sacral nerves, and from the hypogastric and ovarian sympathetic plexuses. The lymphatics from the ovaries end in the retro-peritoneal glands, in front of the aorta.

2. *The Fallopian Tubes, or Oviducts.*—The Fallopian tubes extend from either uterine cornu outwards between the layers, and at the upper fold of, the broad ligament. They measure from 4 to $4\frac{1}{2}$ inches in length. They are divided into three parts for the sake of description :—

(1) The interstitial part, which lies in the uterine wall.

(2) The ampullar part, *i. e.* the distal extremity of the tube. It is the widest portion of the tube, and ends in a number of fringe-like terminations.

(3) The isthmus, or the intermediate part which connects the ampulla and the interstitial part of the tube.

The tube is covered for three fifths of its circumference by the peritoneum which forms the broad ligament, the remaining two fifths lying between the layers of the broad ligament. The blood-supply of the tube is received partly from the uterine, partly from the ovarian arteries, and returns through the uterine and ovarian veins. The nerve-supply and the lymphatic system are the same as those of the ovary.

3. *The Uterus.*—The uterus normally lies almost horizontally within the pelvis, its anterior surface resting upon the bladder. The non-impregnated uterus is about 3 inches in length, $1\frac{1}{2}$ inches in its greatest width, and 1 inch in thickness. It consists of two parts, the body and the cervix, which are united by the isthmus at the level of the os internum. Almost the entire body is covered with peritoneum, except at the sides, whence it is reflected to form the broad ligament. Below, the peritoneum is reflected on to the bladder, at the level of the os internum. On the posterior wall the peritoneum covers the uterus as far as the middle of the cervix, whence it is continued on to the posterior vaginal wall. The uterine wall consists of three coats,—a mucous or internal coat, a muscular or middle coat, and a peritoneal or external coat. The mucous coat is traversed by numerous tubular and branching glands, which are lined by columnar ciliated epithelium. The muscular coat consists of bundles of unstriped muscle-fibres, some of which run longitudinally, others circularly. The blood-supply of the uterus is received mainly from the uterine artery, a branch of the internal iliac, but also to a less extent from the ovarian artery. The uterine veins form a plexus in the broad ligament, and then terminate in

the internal iliac vein.   The nerve-supply of the
uterus is the same as that of the ovaries and tubes.
The lymphatics of the body end in the retro-
peritoneal glands, those from the cervix in the
internal iliac glands, near the point of bifurcation of
the common iliac artery.

4. *The Vagina.*—The vagina is the canal connecting
the uterus with the external organs of generation.
It is lined with a mucous membrane which is covered
with squamous epithelium, and is extremely dilatable.   The vagina contains no glands.   Normally
its anterior wall measures about 2½ inches in length,
its posterior wall about 3½ inches.

The external organs of generation comprise the
mons Veneris, the labia majora and minora, and the
clitoris.   They do not call for any special remarks.

The Ovum.—The ovum is composed of the following parts :—

        (1) Placenta.
        (2) Chorion.
        (3) Amnion.
        (4) Umbilical cord.
        (5) Liquor amnii.
        (6) Fœtus.

(1) The placenta is formed in part from the
chorion and its villi, in part from the decidua
serotina.   It is thus partly fœtal and partly maternal.
In shape it is almost circular, with a diameter of
from 7 to 8 inches.   It is about 1¼ inches thick at
the centre, and weighs about one pound.

(2) and (3) The chorion and the amnion are
respectively the outer and inner layers of the fœtal
membranes.   They are purely fœtal in formation.

(4) The umbilical cord is the means by which the foetal blood is brought to and from the placenta. It consists of two umbilical arteries and one umbilical vein. The arteries convey deoxygenated blood to the placenta, the vein returns reoxygenated blood to the child. Surrounding these vessels lies the Whartonian jelly, while the entire cord is covered by amnion. There are also found in the cord the remains of the allantois and of the vitelline duct. The normal length of the cord is about 22 inches, but it has been found to vary between 6 and 64 inches.

(5) The liquor amnii is the fluid which fills the amniotic sac, and in which the foetus floats. Its normal quantity is from 2 to 5 pints, but as much as 20 pints have been recorded. It is formed principally by transudation from the blood-vessels of the mother, also by the excretion of the foetal skin and kidneys, and by transudation from the placenta and umbilical cord (Winckel). It prevents undue pressure upon the foetus or umbilical cord, it separates the layers of amnion and so prevents their adhering, and during labour it helps to dilate the cervix.

(6) The following table, taken from 'Dührssen's Midwifery,' shows the length of the foetus in centimetres, at the end of the different months. At the end of each month up to the 5th, the length of the foetus in centimetres, is equal to the square of the number of the month. After the fifth month its length is obtained by multiplying the number of the month by five.

Number of Month.		Length of Fœtus in Centimetres.	
1 .	1 × 1 =	1	($\frac{2}{5}$ inch)
2 . .	2 × 2 =	4	(1$\frac{3}{5}$ inches).
3 . .	3 × 3 =	9	(3$\frac{3}{5}$,,).
4 . .	4 × 4 =	16	(6$\frac{2}{5}$,,).
5 . .	5 × 5 =	25	(10 ,,).
6 . .	6 × 5 =	30	(12 ,,).
7 . .	7 × 5 =	35	(14 ,,).
8 . .	8 × 5 =	40	(16 ,,).
9 . .	9 × 5 =	45	(18 ,,).
10 . .	10 × 5 =	50	(20 ,,).

The Fœtal Skull.—The following are the important diameters of the fœtal skull :—

Sub-occipito-bregmatic diameter = 3$\frac{3}{4}$ inches.
Cervico-bregmatic   ,,   = 3$\frac{3}{4}$,,
Occipito-mental   ,,   = 5$\frac{1}{4}$,,
Occipito-frontal   ,,   = 4$\frac{4}{5}$,,
Bi-parietal   ,,   = 3$\frac{3}{4}$,,

The sub-occipito-bregmatic diameter, *i. e.* the distance between the bregma or large fontanelle, and a point just below the occipital prominence, is the greatest diameter of the head that has to pass the brim in a vertex presentation. The cervico-bregmatic diameter, *i. e.* the distance between the bregma and a point representing the junction of the neck and chin, is the greatest diameter that has to pass the brim in a face presentation. The occipito-mental diameter, *i. e.* the distance between the chin and the most distant part of the occiput, is the greatest diameter that has to pass the brim in a brow presentation.

The sutures are the lines of separation between

the bones of the cranium. They consist of the lambdoidal, between the occipital and the parietal bones ; the sagittal, between the parietal bones ; the coronal, between the frontal and the parietal bones ; and the frontal, between the two lateral portions of the frontal bone.

The fontanelles are the angular spaces formed by the intersection of the various sutures ; they are four in number :—

(1) The anterior fontanelle, the large fontanelle, or the bregma, is situated at the junction of the sagittal, coronal, and frontal sutures. It is lozenge-shaped, and four sutures can be felt meeting to form it.

(2) The posterior fontanelle or the small fontanelle is situated at the junction of the lambdoidal and sagittal sutures. It is triangular, and three sutures can be felt meeting to form it.

(3 and 4) The two lateral fontanelles are situated one at either end of the coronal suture.

The different fontanelles can be distinguished when making a vaginal examination, not by their size or shape, but by the number of sutures which meet to form them.

CHAPTER III.

THE DIAGNOSIS OF PREGNANCY.

The Diagnosis of Pregnancy — The Certain, Probable, and Doubtful Signs of Pregnancy—Estimation of the Date of Pregnancy—Is the Fœtus Alive or Dead?

ONE of the most important questions that comes before the obstetric physician, is the task of diagnosing the existence or non-existence of pregnancy. The diagnosis may be all-important, and the result of a mistake, disastrous. The physician who undertakes the consideration of the question should always remember that, though the evidence may be tolerably certain so far as he is concerned, still his diagnosis must be guarded unless absolute certainty dictate it.

The diagnosis is based on certain subjective and objective symptoms.

Amongst the former are classed the cessation of the menses, morning sickness, the movements of the fœtus, salivation, and longings or pica. These in themselves are of slight importance. The patient may wilfully deceive us, or be herself deceived. But when we consider them in conjunction with the objective symptoms, and when we find that the one confirms the other, then they become of value.

And now to consider the objective symptoms. I shall presume that there is no difficulty in the way of a full, thorough, and sufficient examination of the patient. This being so, it is best to examine her in the routine manner adopted in disease.

The face in some cases shows excessive pigmentation occurring at the sides of the nose, under the eyelids, and about the upper lip. *The breasts* come next; they appear enlarged, and a change is noticed in the areola, viz. a secondary areola is more or less well marked. This is a deposit of pigment round the nipple, and varies in depth of colour according to whether the patient be fair or dark. Upon it are seen little nodules—Montgomery's tubercles. By palpation the breasts are discovered to be firmer and more knotty than in non-pregnant women. If they be squeezed, colostrum may exude from the nipple. These changes have also been noticed in cases of myomata of the uterus and ovarian tumours.

Then inspect the abdomen. It is enlarged in correspondence to the period of pregnancy; and as a result of the stretching of the abdominal walls the *striæ gravidarum* appear. They are whitish or bluish lines radiating upwards from the mons veneris, and are due to a loss of elasticity in the cutis vera and the rete Malpighii, consequent on the stretching caused by the enlarging uterus (Winckel). The abdomen also may be more or less pigmented.

By *percussion* we map out the size of the abdominal tumour, and determine whether it be dull or resonant. By this means flatulence and phantom tumours may be excluded.

By *palpation* a tumour is felt if the fourth month be passed. Its size can be determined, and its consistency, the regularity of its surface, and the

2

irregularity of its contents.   In pregnancy the enlarged uterus feels smooth and ovoid, and irregularities
in its contents can be felt, viz. the fœtal parts.   The
fœtus can be moved about between the two hands,—
that is, external ballottement can be obtained.   As
we examine, we notice that the uterus becomes hard
from time to time, *i. e.* it contracts.   There is no
pain accompanying these contractions.

By *auscultation* over the abdomen of a pregnant
woman several different sounds can be heard :—

(1) First and most important is *the fœtal heart.*
It is heard from the sixteenth week onward, beats at
the rate of 130 to 150 per minute, and sounds like
the ticking of a watch.

(2) *The uterine souffle.*   It is a blowing sound
produced in the ascending branches of the uterine
arteries ; it is heard more plainly over some parts of
the uterus than others, and of course is synchronous
with the mother's pulse.

(3) *The funic* or *umbilical souffle.*   It is produced
in the vessels of the cord, probably in the umbilical
vein.   It is synchronous with the fœtal heart, and
generally is caused by the cord being twisted round
the child, or by its being compressed beneath the
stethoscope.   Its presence is usually of bad import
for the child.

(4) *The maternal heart-sounds.*   If they be rapid
they may be mistaken for the heart-sounds of the
child, to avoid which the finger should always be
placed on the mother's pulse whilst auscultating the
fœtal heart.

(5) *Respiratory murmur of the mother.*

(6) *Movements of the child.*

(7) *Friction between uterus and abdominal wall.*

(8) *Crepitating noises* due to air in uterus or abdominal walls.

(9) *The muscular susurrus,* that is the note given out by contracting muscle-fibre.

The vulvar and vaginal mucous membrane is the next thing to inspect. It becomes of a bluish-purple colour, due to venous stasis. This is Jacquemin's and Spiegelberg's sign of pregnancy. It is noticed also in uterine myomata and ovarian tumours, when they attain any considerable size. But in the case of pregnancy it occurs with a smaller uterine enlargement than in the case of myomata.

Vaginal examination is next made, and it is upon the information it gives that we entirely rely for the diagnosis of pregnancy in the early months. First examine the cervix, its length and consistency. As pregnancy advances it becomes apparently shorter than normal, and even in the earlier months it is very much softer. Then try to obtain internal ballottement. . It can be got by passing the fingers into the anterior fornix and pressing suddenly upwards against the uterus. Keep the fingers in the same position, and if the case be suitable the displaced foetus will be felt to fall back upon them, causing a slight sensation of shock. The occurrence of this phenomenon depends on two factors : first, that the foetus be large enough to be felt ; secondly, that it be sufficiently moveable in the liquor amnii to be easily displaced. Both these factors are present during the fourth and fifth months. This sensation of ballottement can possibly be simulated by two other conditions—a pedunculated myoma floating in ascitic fluid, and a large calculus lying in a distended bladder.

Next palpate the ureters; they hypertrophy during pregnancy. To find them palpate the back of the symphysis with the finger in the vagina, and then, starting above at one side of the joint, draw the finger downwards and slightly outwards along the back of the pubes. The ureter, which here lies between the anterior vaginal and the posterior bladder wall, is displaced forwards against the pubis, and is felt to slip from under the finger. It will require some practice to be able to tell if it be enlarged or not. If it be felt at all by the student it is probably hypertrophied, as it is difficult to feel a non-hypertrophied ureter. While we are palpating the ureter the increased pulsation in the lateral fornices is also noticed.

Now attempt to get Hegar's sign of pregnancy— the softening of the lower uterine segment. It is

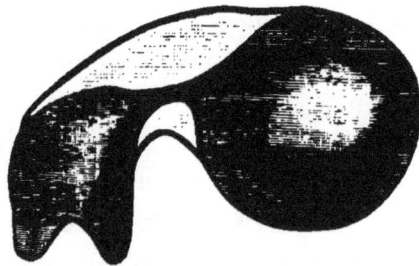

Fig. 1.—Hegar's sign of pregnancy. The heavy outline represents the *actual* shape of the uterus; the shaded portion represents its *apparent* shape as ascertained by recto-vaginal examination. ('The Norris Text-book of Obstetrics.')

best obtained by passing the thumb into the vagina and one finger into the rectum, and then pressing the fundus downwards with the other hand on the abdominal wall, so that the lower uterine segment can be grasped between the finger and thumb. If pregnancy be present, the whole lower uterine segment is so softened that there seems to be no connection between the fundus and the cervix. At the same time the fundus gives the impression to the finger that it is globular (*v.* Fig. 1). This is a reliable sign, very constant and very characteristic. It can be obtained from the second month onwards, but may possibly be obtained in a non-pregnant uterus.

Now let us consider the value of these different signs. They can be divided into doubtful, probable, and certain signs, and can be classified accordingly.

Certain.	Probable.	Doubtful.
The fœtal parts.	Breast changes.	Nausea.
The fœtal heart.	Internal ballottement.	Salivation.
Movements of fœtus when felt by doctor.	Blue colour of vagina.	Pigmentation.
	Increased pulsation in lateral fornix.	Longings.
Funic souffle.	Softening of lower uterine segment.	Cessation of the menses.
	Enlargement of the uterus.	Enlargement of abdomen.
	Hypertrophy of the ureters.	

In default of certain signs, a diagnosis can be made by noting a correspondence between the subjective and objective symptoms. For instance, if the duration of amenorrhœa correspond with the size of the uterus, or if the date of quickening correspond with either of them, then we have a very reliable clue to the situation. The diagnosis

has to be made in the early months from any condition which may give rise to enlargement of the
uterus, as endometritis, or small myomata. The
menstrual history will then usually suffice. Also
from acquired amenorrhœa, due to anæmia, phthisis,
change of conditions of life, &c. This class of cases
is much more difficult to diagnose. A certain
diagnosis can only be arrived at in course of time.
In the later months a diagnosis has to be made from
ovarian and uterine tumours, particularly myomata,
from ascites, flatulence, phantom tumours, &c. The
menstrual history, the time occupied by the growth
of the tumour, the absence of fœtal parts, and
the possibility in some cases of separating the
tumour from the uterus, will usually suffice to make
the diagnosis. In *pseudo-cyesis*, the abdomen is
resonant, and if a whiff of chloroform be administered to the patient, the tumour disappears.

When a diagnosis has been made of the existence
of pregnancy, we have to decide how far pregnancy
has advanced. This can be accomplished by various
methods, none of them, unfortunately, being very
exact. The first way that naturally occurs to us
is to count the weeks that have elapsed since menstruation ceased. This method, although uncertain,
will usually bring us within a fortnight of the true
age, if the woman's history be correct. We can
confirm this by inquiring the date at which quickening occurred, especially in multiparæ, who are
naturally more skilled in detecting it. It usually
takes place about the eighteenth week, but here
again there may be an error of about a fortnight,
too much or too little.

Much more reliable than either of these methods
is the information given by the height of the uterus.
If the pelvis of the patient, and the size of the
uterus, be normal, then by noting the height of the
fundus we can tell at once in what month of pregnancy she is. This can best be described in tabular
form.

At the end of 2nd month the uterus is the size of a large orange.
  ,, 3rd ,,   ,,   ,, fœtal head at
                      term.
  ,, 4th ,, the fundus is at level of top of sym-
                  physis.
  ,, 5th ,,   ,, midway between sym-
                  physis and umbilicus.
  ,, 6th ,,   ., at umbilicus.
  ,, 7th ,,   ,, three fingers' breadth
                  above umbilicus.
  ,, 8th ,,   ,, halfway between umbi-
                  licus and ensiform car-
                  tilage.
  ,, 9th ,,   ,, up to ensiform cartilage.
  ,, 10th ,,   ,, same as at 8th month.

By the various methods which I have described
the age of pregnancy can be estimated very exactly,
and the chance of errors in the patient's history
eliminated. And now by comparing the results
obtained by these various methods, we are able to
predict the day of confinement.

Assuming that the height of the uterus tends to
prove that the menstrual history may be relied on,
we can tell approximately the date of delivery by
the method of Naegele or Matthews Duncan. Pregnancy is usually divided into ten menstrual periods

of four weeks each, that is 280 days. Naegele
counted from the first day of the last menstruation.
He subtracted three months from that, and then
added seven days, or in leap year six days if February
were included in the time. For instance, if the
patient began to menstruate on July 1st, count back
three months, to April 1st; then add seven days, to
April 8th; count forward a year, and the result will
be the date of delivery. The method of Matthews
Duncan is slightly different. He counted from the
last day of last menstruation, and added on nine
months and three days to it. If the menstruation
which began on July 1st ended on July 5th, then
nine months and three days added on brings the date
to April 8th again.

Reckoning from the date of *quickening, i. e.* the
day on which the mother first feels the movements
of the fœtus, and supposing quickening to occur at
the eighteenth week, by adding on twenty-two weeks
we get the required date. The date thus found must
not be considered absolute; it is the centre of a
fortnight during which delivery will probably occur.

The last question to be decided is, whether the
fœtus be alive or dead. Of course, the fact that
pregnancy is continuing is usually an indication that
the fœtus is alive. A dead child usually induces
labour, but sometimes it may be retained in the
uterus. If the patient be past the sixth month, and
still no heart can be heard on the most careful auscultation, the fœtus is probably dead. If the woman
have felt the child frequently, and one day noticed
unusually active movements, and after that there is
a complete cessation of movement, the fœtus is

probably dead.   Lastly, if in conjunction with these symptoms we find that a uterus, which steadily increased in size up to a certain day, has ceased to increase any further, and rather is diminishing in size, the diagnosis is complete.

The woman's symptoms are also of importance. If the child have been dead for any length of time, she begins to lose her appetite, and to become thinner and weaker. She complains of a disagreeable taste in her mouth, and her face assumes a yellowish tinge. Then on making a vaginal examination it may be possible to feel the cranial bones loose and moveable under the skin. If the membranes have ruptured the fœtus decomposes, and a fœtid discharge comes away from the vagina.

CHAPTER IV.

SOME DISEASES OF PREGNANCY.

Morning Sickness and Hyperemesis — Anæmia — Hydræmia—
Varicose Veins—Hæmorrhoids — Salivation — Pyrosis—Pruritus Vulvæ—Neuralgia.

THERE are certain mild disorders which occur during pregnancy which are so constant in their occurrence that they may almost be called physiological.   These are due partly to the pressure of the growing uterus, partly to changes in the blood and in the activity of the nervous centres.

Morning Sickness.—The commonest of these disorders is, perhaps, the nausea or vomiting of the early months,—the so-called morning sickness. This is reflex in origin, and is ascribed by Rheinstadter to the passive movements of an enlarged uterus against the intestines.   It is more likely to be due to the hyper-sensitive condition of the patient. The severity of the condition varies very much in different subjects.   In some it is little more than a slight sense of nausea, whilst in other cases— happily rare—it may reach such a pitch, that the gravest result is to be apprehended.   It is then known as hyperemesis, and becomes one of the severest diseases of pregnancy.   In its mild form morning sickness requires little treatment ; the

regulation of the bowels is generally sufficient.
A cup of hot water the first thing on awaking,
or a very light breakfast in bed at 7.30 a.m. or
8 a.m., consisting of a cup of tea and a small
piece of dry toast, will usually overcome any
tendency to vomit. If this be not enough, the
administration of bicarbonate of soda, subnitrate of
bismuth, aromatic spirits of ammonia, or of a pill
containing calomel $\frac{1}{4}$ gr. and Pulv. Ipecac. $\frac{1}{4}$ gr. may
be tried. I have special faith in hot water, or
in an effervescing mixture containing hydrocyanic
acid.

Hyperemesis Gravidarum.—The severer form of
vomiting, namely, hyperemesis, though certainly
not a physiological disorder, may be mentioned here.
It is a most serious disease. The patient vomits so
constantly, that no food can be retained in the
stomach. She is reduced to a skeleton, and
unless the vomiting can be checked, death follows.
The first essential in the treatment is absolute rest
in bed, her room being kept dark and warm. If
this fail to check the sickness, all food by the
stomach must be prohibited, and rectal alimentation
substituted. Enemata of beef juice, brandy, pep-
tonised milk, egg, are to be used, and sedatives
may be given in the same manner. Bromide of
potassium, chloral, codeia, Tr. Opii, are all recommended. The patient is allowed to suck ice to
relieve her thirst, or to take small sips of cham-
pagne or hot water.

Attention should be called to the local lesion that
often accompanies these manifestations, namely,
erosion of the cervix. Milder forms of vomiting have

been entirely checked by treating it, and especially
by painting the cervix frequently with cocaine.

Another form of treatment is recommended by
Dr Copeman, who states that he has cured many
cases by dilatation of the cervical canal, and detachment of the membranes round the internal os.
His method is to pass the finger in through the os
internum, and keep it in position for a few moments.
The statistics of this method are so favorable that
it certainly deserves a trial. It should be combined
with the treatment of any erosion that may be
present. If done aseptically, it at any rate does no
harm, and may be tried before resorting to artificial
abortion. If none of these methods succeed, then
nothing remains but the induction of abortion. The
worst of this treatment is, that, being an extreme
measure, it is put off in most cases so long that
when it is accomplished the patient is too far gone
for recovery.

The prognosis of hyperemesis is bad. Joulin has
reported 121 cases with 49 deaths, or something over
40 per cent.

Anæmia.—Normally the number of red blood-corpuscles is increased during pregnancy, but sometimes
the opposite is the case. The commonest causes of
this condition are bad food, bad digestion, insufficient
exercise in the open air. The treatment is largely
prophylactic. If any active measures are necessary,
iron in some form is administered. Lusk especially
recommends Ferrum Redactum in 3-grain doses.
The bowels must be regulated by the use of purgatives, and for this purpose tabloids containing aloin in
conjunction with ferrous carbonate are of great use.

Hydræmia.—This may be considered in connection with anæmia, as they often occur together. When hydræmia occurs, it not infrequently causes œdema of the lower extremities and vulva. If the possibility of renal disease be excluded, this condition is of slight importance. If the œdema of the labia be excessive, they may offer an obstruction to delivery, or even in some cases may become gangrenous. Unless either of these terminations be feared, the condition requires little treatment. If the œdema be moderate, rest in the recumbent position and the application of lead lotion will relieve it. In cases of enormous distension of the labia it may be necessary to puncture them. The danger of this proceeding is that suppuration may occur, consequently it should only be resorted to if absolutely necessary, and then every possible precaution must be taken to prevent suppuration.

Varicose veins frequently form in the later months of pregnancy. The treatment must be palliative, as an operation is by many considered to be contra-indicated, unless absolutely necessary. An elastic bandage applied to the affected limb is usually sufficient. If the veins still continue to increase in size and threaten to rupture, the patient should be kept in a recumbent position.

Hæmorrhoids are also a very common trouble at the end of pregnancy. Usually they disappear a short time after delivery, but sometimes they persist, and give rise to such annoyance as to necessitate their removal. During pregnancy, the only treatment which we can adopt is to keep the motions soft and regular, and to use some soothing application, as Ung.

Galli c̄ Opio. Tr. Arnicæ in water, two drachms to the ounce, Tr. Opii in water, are also useful.

Salivation.—This may be extremely troublesome, but it is a rare affection. Desire the patient to wash out the mouth with an astringent solution, or administer internally two- to four-minim doses of Liquor Atropinæ Sulphatis.

Pyrosis.—This condition depends upon bad digestion of food, or constipation. If the cause be treated it will cease. The common Mistura Ferri Sulphatis of hospital pharmacopœias is one of the best mixtures that can be used. It is as follows :—

> ℞　Mag. Sulph. ʒj.
> 　　Acidi Sulph. Dil. ʒiss.
> 　　Ferri Sulph. gr. viij.
> 　　Aquæ Menth. Pip. ad ʒviij.
> Sig. ʒj ter in die.

If the patient cannot take this, a pill, containing a third of a grain of calomel, three times a day is an excellent laxative, can always be taken, and is usually sufficient.

Pruritus Vulvæ.—This is a most distressing condition. It is usually caused by vaginal discharge, but may also be due to diabetes and parasites. If vaginal discharge be the cause, the great point in the treatment is absolute cleanliness. The patient should bathe the parts twice or thrice daily with warm water, in which some mild astringent is dissolved, as borax, very dilute sulphate of copper solution, or pyroligneous acid. Vaginal douches at the same time may be used, such as pyroligneous acid 1 in 4, boracic acid (saturated solution) or carbolic acid 1 in 80.

If erosion of the cervix be the cause of the discharge it should be touched with pure carbolic acid, or a solution of sulphate of copper, thirty grains to the ounce, or pyroligneous acid applied every couple of days. If these manipulations be performed with gentleness there is no fear of inducing labour. At the same time the pruritus itself may be relieved by applying some soothing ointment, as Ung. Oxidi Zinci. This acts by preventing the discharge from coming in contact with the skin, while at the same time it relieves the irritation.

Neuralgia.—Various neuralgic affections are common in pregnancy, particularly involving the regions supplied by fifth cranial nerve. Local application of warmth, or of camphor or chloroform liniment, will usually relieve the pain. Quinine, bromide of potash, or hyoscyamus, especially the first, are also sometimes of use.

CHAPTER V.

LABOUR AND ITS MANAGEMENT.

The Causes of Labour—Diagnosis of Labour—Preparations for Labour—Care of the Nipples—Method of preparing the Obstetric Couch—Obstetrical Armamentarium.

Causes of Labour.—What the factors are which cause a pregnant uterus to contract at the tenth menstrual period after conception, and to expel its contents, is little known. So far our views are but the results of conjecture. We know that certain changes occur during pregnancy, and we infer a consequence for them. What these changes are I shall state in a few words.

First.—The uterus and ovum increase in size during pregnancy. In the earlier months the uterus grows more rapidly than the ovum; but in the later months the ovum grows faster than the uterus. Hence it comes about, that, towards the end of pregnancy, the growing ovum tends to become too large for, and so to exert a distending pressure upon, the uterus.

Secondly.—During the entire period of pregnancy the uterus shows a certain amount of irritability and a tendency to contract intermittently. This irritability is especially marked at the menstrual period, and becomes more marked with each successive period. In the later months it shows itself by the painless contractions of the uterus—an intermittent series of contractions which can be felt by laying the hand upon the uterus.

Thirdly.—The cells of the decidua vera undergo a fatty degeneration toward the end of pregnancy.

Fourthly.—Large giant-cells appear in the decidua serotina. After the eighth month they invade the uterine sinuses, cause coagulation of the blood in them, and give rise to the growth of young connective tissue, which still further tends to obliterate the supplying vessels. As each vessel is obliterated the amount of venous blood in the inter-villous spaces is increased.

Fifthly.—There is a motor centre for uterine contraction in the medulla oblongata. Carbonic acid gas in the blood stimulates it. The rapidly growing fœtus daily extracts more oxygen from the maternal blood, and returns, instead of it, a daily increasing quantity of carbonic acid gas. Direct stimuli applied to the uterus will also cause contraction by means of a reflex centre in the spinal cord, and the CO_2 in the blood here again will act as the stimulus.

These are the facts which we know. What conclusions may we draw from them? We have a series of changes occurring in the uterus and its contents, changes that become more marked daily as pregnancy advances. The uterus is becoming more and more distended by the growing ovum, and one day it must become over-distended. The irritability of the muscular fibres is increasing daily, and is tending to cause a separation between

3

the ovum and the uterus. The fatty degeneration
of the decidual cells is paving the way for this
separation to occur more easily. Once it occurs, the
ovum becomes a foreign body, and is expelled accordingly. The giant-cells and young connective
tissue cause a venous condition of the blood, which
furnishes an ever-increasing peripheral stimulus to
the centres in the cord. The growth of the fœtus,
daily abstracting more oxygen from the mother,
causes a daily increasing excess of CO_2 in the
maternal blood, so furnishing the necessary stimulus
for the medulla. All these are predisposing factors ;
and, ever increasing, apparently come to a climax at
the tenth menstrual period. An exciting cause is furnished by some sudden movement—straining at stool,
a violent cough, or the like ; the period of unstable
equilibrium comes to an end, and labour commences.

Is the patient in labour? This is a difficult
question to decide in the early stages of labour.
Later, when the patient is having strong labour
pains, there is no difficulty in making a diagnosis.
Also, in many cases we can say definitely that
the patient is not in labour at the moment of examination, but still we are unable to say that she will
not be in labour within the next hour. And frequently cases are sent out of hospital in the morning
obviously not in labour, who return the same
evening, perhaps, as a case of street delivery.

To decide the question a careful examination
of the patient must be made. First palpate the
abdomen, and notice if the presenting part be fixed,
and if the uterus contract intermittently. The fixity
of the head is a reliable guide in multiparæ, but is

of no value in primiparæ. In the former the head
does not become fixed until the onset of labour ; in
the latter it is fixed for the last three or four
weeks of pregnancy. There are a few conditions
which by their presence prevent the head from
becoming fixed at its proper time, and so cause exceptions to this rule. These are ;—contracted pelvis,
hydramnios, hydrocephalic head, face or brow presentation, twins, placenta prævia. In the absence
of these exceptions, the rule given above may be
relied on ; and, if we find the head not fixed in the
case of a multipara, probably she is not in labour,
and *vice versâ*. If uterine contractions can be felt,
find out if the patient complains of pain during
them,—that is if the contractions are painful or
painless. The presence of painless contractions may
be taken as a sure indication that she is not in
labour, of painful contractions that she is.

A vaginal examination must next be made, with a
view to discovering whether the os be dilating or not.
If it be only slightly dilated, the patient may not be
in labour. One frequently meets with an os the size
of sixpence three or four days before labour has set
in. If, however, the os be tightly closed, then she is
not in labour.

There are two other points of slight importance.
One is the so-called " show," a blood-stained mucous
discharge which comes from the vagina for one or
two days before labour sets in. Another is the onset
of false pains which occur in various parts of the
abdomen, but not in the back.

Preparation for Labour.—During the last fortnight
of pregnancy the patient should be taught to pay

particular attention to certain points. She should
have warm baths daily; and her bowels should he
regulated, so as to avoid the constipation which
occurs, especially, towards the end of pregnancy. As
soon as the premonitory symptoms of labour set in, a
good purgative must be administered—castor oil (two
ounces), sulphate of magnesia (half an ounce), or
cascara sagrada (two drachms), followed after twelve
hours by an enema. Another enema must be given
as soon as labour has well set in.

The due care of the breasts is a most important
point, especially in primiparæ. The physician should
examine the nipples to ascertain if they be of a
shape suited for nursing. If they be at all depressed,
the mother must be taught to draw them out gently
with her fingers several times a day, taking care
not to use undue force, and to have perfectly
clean fingers. Too violent attempts at forming the
nipple, especially when they are made by an unskilled
nurse, often result in causing slight lacerations in the
delicate skin. Then, if the fingers or nipples be
dirty, the cracks become infected, and mastitis may
follow. In addition to forming the nipple the patient
must bathe them, a couple of times daily, with some
lotion that will harden the skin. Otherwise a strong
healthy child will cause the greatest pain whilst
nursing. The best lotion to use is alcohol in some
form—eau de Cologne, whisky, or common methy-
lated spirits. Begin with a weak solution and
gradually increase the strength. Pure whisky may
be used, but eau de Cologne must be diluted by
adding an equal volume of water.

In the case of a primipara it is always necessary

to instruct her as to what she requires to have in readiness for her delivery. The following list will be found fairly complete ;—two mackintoshes, one large enough to cover the bed, the other about one third that size ; four binders 1¼ yards long, and 18 inches wide ; half a dozen packets of sanitary towels; half an ounce of surgical pins ; one skein of glazed linen thread ; one pound of absorbent gamgee tissue to use as sponges.

It is also of importance that the physician should know how the obstetric bed is made. In making it we require to combine comfort with cleanliness and convenience. The patient must not be, as our grand-mothers were, lost at the bottom of a large feather bed; neither must she be forced to lie on a hard board. A good, firm, hair mattress will meet every requirement ; it should have boards beneath it in-stead of springs. The bed should be about two feet in height. It is made in the following man-ner, from below upwards : — (1) the mattress, (2) the large mackintosh ; (3) an under blanket ; (4) a sheet and bolster ; (5) the small mackintosh wrapped up in the drawsheet ; (6) a pillow ; (7) a top sheet with the requisite number of blankets. There should be a piece of oil-cloth or mackintosh hanging down as a valance at the side of the bed, in order to protect it.

The other essentials in the room are,—a fire, unless the weather be extremely warm ; and it should, if possible, be one on which a kettle can be boiled ; a large jug thoroughly scoured inside and outside, to hold about one and a half gallons ; a stand of some kind on which it can be placed, and which will

raise it about two feet above the patient's bed; two
additional jugs, one for cold and one for hot water;
four basins if possible, but three are sufficient; an
abundant supply of boiling and of cold water.

And now of what must the doctor's armament
consist ? I shall give first the things necessary for a
perfectly normal case, and then a list of everything
that will be required for any operation short of
abdominal section. For a perfectly normal case he
requires :—

    (1) Corrosive sublimate tabloids.
    (2) A bottle of creolin or lysol.
    (3) A piece of carbolic soap.
    (4) A good nail-brush.
    (5) A metal catheter.
    (6) A Higginson's syringe for administering
        enemata.
    (7) A bottle of some preparation of ergot.

I particularly omit in this list any mention of a
douche for douching the vagina or uterus. In
normal uncomplicated labour douching of any kind
is unnecessary.

In order to be prepared for any obstetrical
emergency except abdominal section, the following in
addition are required :—

(1) A syphon douche and glass nozzle. The best
kind consists of a plain rubber tube, about six feet
in length, without valves of any kind. At one end
it has got a sinker which keeps it immersed in the
fluid used ; a little further up the tube is encased in
a moveable horseshoe-shaped piece of vulcanite,
which fits on the top of the jug and prevents the
tube from kinking. Halfway down the tube there is

a ball-shaped expansion, and a little further on there
may or may not be a tap. It is completed at the
other end by a glass nozzle. To use the douche the
sinker is immersed in the fluid, and the vulcanite
support adapted to the depth of the jug, which is
then placed upon the stand. Now compress the ball
with one hand, and having done so squeeze the tube
between the nozzle and the ball. By this means

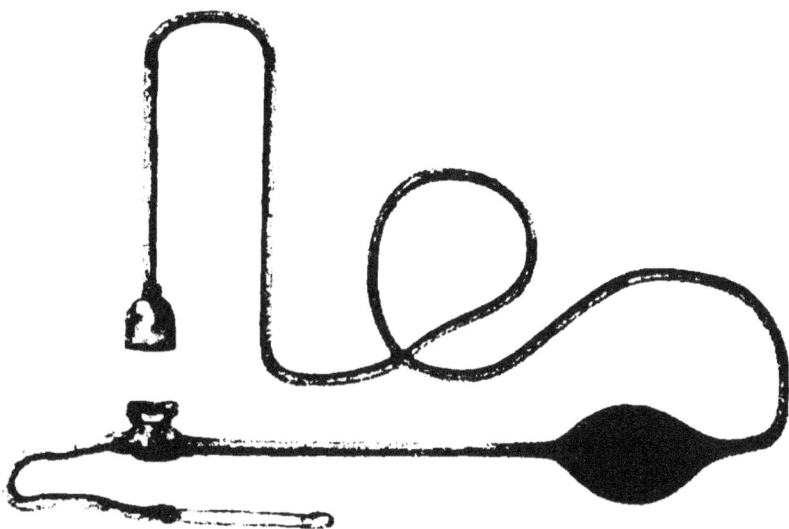

FIG. 2.—Syphon douche, as described in the text.

when the ball is released water is drawn into it from
the jug. This usually is sufficient, and the water
will continue to run, upon the principle of a syphon.
If it do not, it is only necessary to repeat the previous
manipulation a second time. (*v.* Fig. 2.)

    (2) An ordinary needle-holder, and a few large
        and a few small half-curved needles.

    (3) An axis-traction forceps.

(4) A silver male catheter, for removing mucus from the child's larynx.

(5) Two Bozeman's uterine catheters, one large and one small.

(6) A perforator—Naegele's or Simpson's.

(7) A Braun's blunt hook for decapitation.'

(8) A cranioclast—Braun's or Winter's modification of Auvard's.

(9) A long narrow forceps for plugging the uterus.

(10) A couple of American bullet forceps.

(11) A posterior speculum.

(12) Two or three curettes, including Martin's and Rheinstadter's.

(13) Asepticised silk and catgut.

(14) Chloroform and an inhaler.

(15) Iodoform gauze for plugging purposes.

(16) A box of absorbent cotton wool for the same purpose.

(17) A couple of gum-elastic catheters, Nos. 10—12, to act as port-fillets or as repositors.

(18) A hypodermic syringe.

(19) The following drugs :—opium, ether, sal volatile, perchloride of iron.

This list may seem to be excessive, and to include more appliances than can be carried in a bag. If the perforation instruments be excluded, as they usually can be, the remainder can with ease be held in an ordinary midwifery bag.

CHAPTER VI.

DIAGNOSIS OF THE PRESENTATION AND POSITION OF THE FŒTUS.

The Attitude of the Fœtus—The Presentation—The Position—
Diagnosis of the Presentation and Position — Abdominal
Palpation—Vaginal Examination—Auscultation.

Attitude.—The attitude which the fœtus assumes
in the uterus is one which reduces it to the smallest
possible size. The head is flexed upon the chest,
the spine curved forward, the upper limbs crossed
on the chest, the thighs flexed on the abdomen, and
the legs on the thighs. It thus assumes an ovoid
shape, having two poles, the pelvic and the cephalic.
The pelvic pole, that is the breech and lower
extremities, is larger than the cephalic pole; and
the distance between the two poles is the longest
diameter of the ovoid.

Presentation.—The presentation of the fœtus is
another matter. It is the term applied to the part
of the fœtus which lies immediately over the os
internum, and which would therefore be the first
part born, if the fœtus were suddenly extruded. Any
part of the fœtal ovoid can present, but, as a
matter of fact, it is in a large majority of cases one
or other of the poles. Of these two poles, the
cephalic presents far the more frequently. Indeed,
out of all full-term labours, 97 per cent. of the
children come head first. It is easy to understand

why the child should present by one or other pole ;
but why there should be such an overwhelming proportion of head presentations is at first difficult to
understand.

In the earlier months of pregnancy the uterus
grows more rapidly than the ovum, and consequently
the fœtus is free to move about and assume any
position. As pregnancy advances the fœtus grows
more rapidly than the uterus ; and as it commences
to occupy the entire uterus, it is guided round until
its long diameter corresponds with the long diameter
of the uterus,—that is, until the fœtus is lying with
one pole at the fundus and the other in the lower
uterine segment. And again, since the breech and
the lower limbs together are more bulky than the
head, and since the uterine cavity is also of an ovoid
shape, whose larger end is uppermost, the pelvic
pole is guided to the fundus, and the head is guided
to the lower uterine segment. If this be so, we
should expect malpositions in cases in which the
uterus has lost its shape. And so hydramnios,
multiple pregnancies, pluriparous uteri, contracted
pelves, uterine tumours, all favour malpositions by
destroying the natural shape of the uterus.

The other hypothesis, that cephalic presentations
are the result of gravity, and that the head falls into
the pelvis because it is heavier than the breech, is
upset by two facts :—

First.—That in women who pass the entire period
of pregnancy in bed lying upon their backs the
head still presents.

Second.—That if a fœtus, tied up in the attitude
in which it normally lies in the uterus, be immersed

in a fluid of the same specific gravity as the liquor amnii, it is the breech which occupies the lowest level in the fluid, and not the head.

Although, theoretically, any part of the fœtus may present, still for practical purposes we recognise five chief presentations. These are in the order of frequency :—

(1) Vertex, occurring in 95·53 per cent. of full-term deliveries.

(2) Breech, occurring in 3·11 per cent.

(3) Face, occurring in 0·6 per cent.

(4) Transverse, occurring in 0·56 per cent.

(5) Brow, occurring in 0·2 per cent.

These different presentations are classified in three groups—normal, natural, unnatural. In the group of normal presentations, vertex presentations alone can be placed. In the group of natural presentations are included any presentations which can deliver themselves as such,—that is to say, face presentations and breech presentations. Vertex presentation should, strictly, also be included in this group, if it were not more convenient to assign a special group to it, and to designate it as normal. In the group of unnatural presentations come the presentations which cannot deliver themselves as such, namely, transverse and brow presentations.

To tabulate the presentations and their frequency :—

Normal	.	Vertex presentations occur in 95·53 per cent. of all cases.				
Natural	. {	Breech	,,	,,	3·11	,,	,,
		Face	,,	,,	0·60	,,	,,
Unnatural	{	Transverse	,,	,,	0·56	,,	,,
		Brow	,,	,,	0·20	,,	,,

Position.—By position we mean the relationship which exists between the dorsal plane of the child and the middle line of the mother, in the case of polar presentations. Transverse presentations are so irregular as regards both the exact presentation and the direction in which the child lies, that no definition could be found to cover correctly the meaning of the word, in them and in polar presentations. Referring then to polar presentations, we usually classify them into two positions, according as the back is to the left or to the right of the middle line of the mother's abdomen. The former is called the first position, the latter the second. Both of these can be, and indeed sometimes are, divided into two subdivisions, according as the back is inclined anteriorly or posteriorly. Thus are got the four positions of Naegele. But inasmuch as the mechanism is almost identical whether the back be anterior or posterior, it seems superfluous to recognise more than two positions. And, indeed, each position might be divided into any indefinite number of sub-positions, seeing that the presenting part may enter the brim in any imaginable diameter of the brim.

Diagnosis of Presentations and Positions.—A diagnosis of the presentation and position of the fœtus can be determined, in any given case, by three methods,—abdominal palpation, vaginal examination, auscultation. Of these, by far the most valuable information can be obtained by means of abdominal palpation. I therefore should like to say a few words upon the subject.

In the first place, what can be learnt by *abdominal*

palpation ? In the second place, how is it performed ? By palpation of the abdomen we can ascertain seven important facts :—

1. *The presence or absence of pregnancy,* at any rate from the seventh month onwards, by feeling a tumour corresponding in size and shape to the uterus, and by feeling fœtal parts within it.

2. *The period of pregnancy.* By mapping out the height of the uterus (*v.* page 23).

3. *The presentation and position of the fœtus.* To palpate a pregnant uterus thoroughly we use four distinct *grips,* or methods of applying the hands. First, place the patient flat upon her back, with her pelvis straight and her legs extended. Then sit down at her right side, about the level of the pelvis and facing her head. Next lay both hands, gently, flat upon the fundus of the uterus, and feel what is lying there. This is called the *fundal grip.* Notice the shape and mobility of the part of the fœtus lying beneath the hands. The hands should be warm ; and we must be careful to avoid undue pressure, as it causes pain, and then the woman contracts her abdominal muscles and renders further palpation impossible. Avoid also lifting the finger-tips off the abdomen—playing the piano on the abdomen,—as this also causes contractions of the recti. Move the fingers and hands gently from place to place without lifting them off.

Having palpated the fundus, move the hands gently downwards until the level of the umbilicus is reached. Then, in the same manner, make the *umbilical grip.* By moving the hands about, the nature of the fœtal parts at that level can be ascertained.

To determine upon which side of the uterus the back of the child lies, lay the hands flat on either side of the uterus and move them synchronously, first to one side, then to the other, making the uterine contents move with them. By this means one notices that there is a greater resistance offered to one hand than to the other. This resistance is usually on the side at which the back is.

The next grip is the *pelvic* or *Pawlic's grip.* This is made with the right hand only. Sink the fingers into the false pelvis over the centre of Poupart's ligament on the left side, and the thumb into the corresponding point on the right, and then approximate them. By this means we discover what is lying in the pelvic brim, and whether it be moveable or fixed. These three grips are usually sufficient to tell all that is required. But if the presenting part have sunk deeply into the brim, then the fourth grip is necessary in order to feel it. To practise this grip two hands are required; and, in place of facing the patient's head, you turn so as to face her feet. Sink the tips of the fingers of the right hand into the true pelvis at one side, and the tips of the fingers of the left hand similarly at the other side. By this means the extent that the presenting part has descended can be estimated.

To explain better the method of palpation, I shall describe a case in which the child is lying in the first vertex position. On making the *fundal grip* a large, hard, round tumour is felt at the fundus, and proceeding from it is felt the back of the child on the left, and perhaps the limbs on the right. There is no groove between the round tumour and the back, and on moving the former it moves *en bloc* with the back. These two points exclude the possibility of the fundal tumour being the head of the child, and so by a process of exclusion it must be the breech. By the *umbilical grip* the back is felt on the left, and on moving the hands laterally the greatest resistance is felt upon the same side. By *Pawlic's grip* another hard round tumour is felt in the pelvic brim. It differs from the tumour at the fundus in that there is a groove running between it and the back—the groove of the neck. This groove, and also the round tumour, lie higher above the symphysis upon the right side than upon the left. If the tumour be not fixed it can be moved about—ballotted—independently of the back. These points distinguish it as the head ; and the fact that the tumour is higher above the symphysis on the right, than on the left, shows that the chin is higher than the occiput, and therefore that the vertex is presenting (*v.* Fig. 3).

The foregoing is the usual manner of performing abdominal palpation ; the special features of the different presentations will be dealt with under the presentations.

4. *The presence of pelvic contraction.* This can be determined at term in some cases. I have mentioned the general rule as to the fixity of the presenting head at term in primiparæ and in multiparæ. In the former it is fixed during the last three or four weeks of pregnancy, in the latter it does not fix until the commencement of labour. There are six conditions which tend to prevent the fixation of the head :—

(1) Contracted pelvis.

(2) Hydramnios.

(3) Multiple pregnancy.

(4) Placenta prævia.

FIG. 3.—Relative position of the chin and occiput in vertex, brow, and face presentations; as ascertained by abdominal palpation (diagrammatic).

(5) Face or brow presentation.

(6) Hydrocephalic head.

In most cases (2) to (5) can be excluded, and

usually also (6). (1) then alone remains, and it is by far the commonest cause of non-fixation of the head. If we meet a case in which the head ballotts freely above the brim at a time at which it should be fixed, pelvic contraction is the first condition to be thought of.

5. *If the patient be in labour.* The diagnosis of labour has already been gone into (*v.* p. 34). The important points are, the presence of true pains or of painless contractions, and the fixity or non-fixity of the presenting part in multiparæ.

6. *The course and progress of labour.* The progress of labour is best determined by noting the descent of the presenting part. In the early stages the height of the chin above the symphysis can be measured in finger-breadths. As labour advances the chin approaches the level of the symphysis, and then sinks below it. The rate of advance can then be determined by the fourth grip.

7. *The indications of threatened rupture of the uterus.* The indications which can be determined by palpation are as follows :—The rising of Bandl's ring upwards on the abdomen. This ring, which marks the line of junction between the contractile upper uterine segment and the non-contractile lower uterine segment, is felt as a depression running across the uterus (*v.* Fig. 31). In normal labour it is not noticed, as it does not rise at all, or only very slightly above the symphysis. In delayed labour, however, Bandl's ring is always rising higher upon the abdomen, according as the muscle-fibres retract and the upper uterine segment thickens. If this ring rise more than 1½ inches above the symphysis, it constitutes one

4

of the signs of threatened rupture of the uterus, and is an indication for immediate delivery. Bandl's ring has to be diagnosed from a distended bladder, as the depression which is found above the latter is not unlike the contraction ring. A distended bladder may usually be recognised by obtaining fluctuation in it, and if a catheter be passed the depression disappears. The standing out of the round ligaments is an indication of danger. They can readily be felt through the abdominal walls like tense cords. The character of the uterine contractions is also of importance, and can be determined by palpation. Normally they should be intermittent, but if labour be unduly prolonged they become continuous or tonic. In certain conditions of the patient abdominal palpation may be impossible;—if she will not allow her abdominal muscles to relax; if the liquor amnii have escaped for a long time, and the uterus be contracted down upon the fœtus; or if there be a great excess of liquor amnii—hydramnios.

The next method of diagnosing the position is *vaginal examination.* By it can be determined;—the nature of the presenting part; the fixity of the presenting part; the condition of the membranes; the size of the os uteri; and the presence of a prolapsed limb or cord. Also, if the presenting part be not fixed, some idea can be obtained as to the size and shape of the pelvis.

The presenting part can be determined by noting its size, shape, and characteristics. A vertex and a breech both feel to be hard, round, smooth tumours; but on the vertex are felt the sutures and fontanelles; on the breech the anus, the tip of the coccyx, and the two tubera ischii. The face, when its features are obscured by a large caput succedaneum, also feels smooth and round. It is recognised by feeling the mouth with the tongue and alveolar ridges, and the supra-orbital ridges. A brow is known by feeling, on one side of the presenting part, the smooth forehead; and on the other, the supra-orbital ridges, and the edges of the orbital cavity. A foot can be distinguished from a hand by feeling the heel; by noting that the line of the tops of the toes is straight, of the tops of the fingers curved; that the thumb can be apposed and opposed, while the great toe cannot. The knee can be distinguished from the elbow by its greater size; by feeling the patellar ligament and the patella, if the knee be not flexed; and especially by feeling the tuberosity of the tibia.

There are no insuperable difficulties in the way of making a vaginal examination; but, unfortunately, there is extreme danger. Very many puerperal women die as a result of septic infection; and, if there were no vaginal examinations, there would be no cases of acute sepsis, in previously healthy women. If, then, vaginal examinations could be entirely abolished, or, at any rate, reduced to a minimum, very many lives would be saved. Let us see how far it can be replaced by abdominal palpation.

If the capabilities of both methods be inquired into, it will be seen, that, while many facts can be determined by abdominal palpation which cannot be determined by vaginal examination, there are very few facts which can be determined by vaginal examination alone. What are these exceptions? The

most important is the diagnosis of prolapse or presentation of the cord. This certainly, as far as we
know at present, cannot be determined by palpation.
It is a most important condition to recognise, and,
therefore, one vaginal examination, at all events,
must be made. The best time to make it is immediately after the rupture of the membranes, as it is
then that the cord prolapses. Another point, that
can be determined by vaginal examination alone, is
the exact size of the os uteri. It, however, is not a
matter of very vital importance, and usually can be
sufficiently nearly ascertained by noting the descent
of the presenting part. All this points to the extreme importance of acquiring skill in practising
abdominal palpation. If we possess it, the number
of vaginal examinations can be very greatly re-
stricted.

The third and last method of diagnosing, or
rather of assisting to diagnose, the position, is
auscultation of the fœtal heart. According to the
presentation and position of the fœtus, the heart is
heard with maximum intensity over one or other part
of the abdomen. Let us imagine the abdomen
divided into quarters by one line drawn vertically,
and another drawn horizontally, through the umbilicus. Then if the head be in the lower uterine
segment, the heart will be heard best below the
transverse line ; and if the head be in the fundus,
above the same line. If the back, in a vertex or
breech presentation, be to the left of the vertical
line, the heart is best heard to the left of the same
line ; if to the right of the line, the heart is heard
to the right. In a face presentation this rule does

not hold good. In that case the heart is usually heard best on the side of the uterus at which the limbs are; and so, if the back be on the left, the heart will be heard on the right, and *vice versâ*.

CHAPTER VII.

NORMAL LABOUR.

Definitions—Stages of Labour—The Phenomena of Labour:
Physiological, Plastic, Mechanical—The Movements of the
Foetus: Descent, Flexion, Internal Rotation, Extension,
External Rotation.

Definition.—Normal labour consists in the child
presenting by its vertex, in the pains coming on, and
following one another, in such a manner, that the
child is born, and everything is over without artificial
aid, within twenty-four hours. This train of events
will happen in about 90 per cent. of all labours.
It is very important, then, to study the phenomena
and management of normal labour, as it is in the
management of it that by far the greater number of
mistakes are made. Frequently by the ignorance
and meddlesomeness of the medical attendant, cases
of normal labour are turned into abnormal ones.

Labour is classified as follows, according to the
time at which it occurs :—

(1) *Abortion,* when it occurs before the formation
   of the placenta, *i. e.* before the commencement of the fourth lunar month.

(2) *Partus Immaturus,* or *miscarriage,* when it
   occurs after the formation of the placenta,
   but before the child is viable, *i. e.* from the

commencement of the fourth to the end of the seventh lunar month.

(3) *Partus Prematurus,* or *premature birth,* if it occur after the child has become viable, but before full time, *i. e.* before the end of the tenth lunar month.

(4) *Partus maturus,* or *full-term birth,* when it occurs at term.

(5) *Partus serotinus,* or *delayed labour,* when it occurs more than forty-one weeks after conception.

In this chapter we are concerned only with *partus maturus,* or full-term birth.

Normal labour is divided into three stages :—

(1) The first stage, or stage of dilatation.

(2) The second stage, or stage of expulsion.

(3) The third stage, or placental stage.

The first stage commences with the onset of true labour pains, and lasts until the full dilatation of the os. Its average length is, in primiparæ, about fifteen hours; in multiparæ, about eleven hours. The second stage commences with the full dilatation of the os, and ends with the expulsion of the child. Its average duration is about two and a half hours in primiparæ and one hour in multiparæ. The third stage commences after the birth of the child, and ends with the expulsion of the placenta. Its length varies greatly according as it ends spontaneously, or is ended artificially.

Phenomena.—The phenomena of labour are usually divided into three groups,—physiological, mechanical, and plastic.

The physiological phenomena include the nature

of the uterine contractions and their effect upon the
ovum and upon the soft parts of the mother. The
so-called 'labour pains' are a series of contractions of
the muscular fibres of the uterus and ligaments,
which occur intermittently, and sweep over the organ
as a peristaltic wave. They act in such a manner
as to cause a diminution in the transverse diameters
of the uterus; and an increase in the longitudinal
diameter, and in the thickness of the walls. The
result of the contraction is also to diminish the
cavity of the uterus, and so to cause pressure upon
the ovum. When a body is acted upon by several
forces, each acting in a different direction, it tends to
move in the direction of least resistance. The ovum,
when pressed upon by the contractions of the uterus,
is driven downwards against the lower uterine segment, and comes to press upon the internal os.
Three factors unite in making the region of the
internal os the area of least resistance. These are :—

(1) The muscle fibres in that part of the uterus
    are less numerous, and so the contractions
    are not so strong as in other parts.
(2) The contractions of the abdominal muscles
    tend to press the ovum downwards.
(3) Gravity also pulls the ovum down in the same
    direction.

While the uterine contractions are at work intermittently diminishing the uterine cavity, there is
another and more persistent change taking place
in the uterine muscle. This is the phenomenon of
retraction of the muscle fibres. To understand it
properly, it must be known that the uterus consists
of two distinct parts or segments,—an upper or

contractile segment, and a lower or non-contractile
segment. The junction between these two segments
is, at the commencement of labour, at, or about,
the internal os. During a pain the fibres which
compose the upper segment not only contract but
retract. By contracting, each fibre diminishes in
length ; while, by retracting, the fibres come actually
to assume new positions, at least, in their relationship to one another ; so that fibres which originally
lay end to end come, after a little time, to lie
parallel. Retraction of the fibres is permanent, so
that they do not return to their original positions
when the contraction is over. The effect of retraction on the uterus is, that, the fibres tend to become drawn upwards towards the fundus; consequently
the contractile portion of the uterus becomes thicker
and shorter and the non-contractile portion becomes
thinner and longer. In other words, the junction
between the upper and lower uterine segments tends
to move upwards towards the fundus. To this
junction the term contraction ring, or ring of Bandl,
is applied. The ring can be felt through the
abdominal walls as a depression, running across the
uterus, in cases in which labour has been strong or
unduly protracted. It is most essential to be able
to recognise this ring, as it gives an absolute indication of the effect of the uterine contractions on the
uterine walls. In normal labours it can seldom be
felt, as it does not rise above the symphysis pubis. If
labour be protracted it rises gradually upwards
(v. Fig. 31). There is said to be threatened rupture
of the thinned lower uterine segment if Bandl's ring
rise more than 1½ inches above the symphysis.

The result of the pressure of the ovum on the internal os, is to cause it to dilate, and it becomes so enlarged as to be practically obliterated; thus the cervical canal becomes continuous with the cavity of the uterus, and the ovum comes to impinge upon the external os. In addition to the pressure of the advancing ovum, other factors unite to cause the dilatation of the cervix. There are, in the lower uterine segment, longitudinal bands of muscle tissue. These, by contracting, draw the cervix upwards over the presenting part, and continue to draw it up until the entire cervix is obliterated, and the vagina, cervical canal, and uterine cavity become continuous one with the other. As soon as this point is reached the os is said to be fully dilated. Another element, in the distension of the cervix, is the serous transudation which takes place into the lymphatic interspaces of the cervical tissue. This is due to a hyperæmia of the vessels, and also to the fact that the return flow through the veins is impeded; and so a transudation takes place which renders the part soft and dilatable. As soon as the cervical canal is obliterated the membranes rupture, the fœtal head advances into the vagina, and descends until it comes to press upon, and so to dilate, the perinæum. The dilatation of the vagina and perinæum is aided, as in the case of the cervix, by serous transudation into their tissues from the blood-vessels.

Plastic Phenomena.—The changes that occur in the shape of the fœtal head, and the formation of the caput succedaneum, are included under this title. The changes in the fœtal head, or the moulding of the head, are due to the pressure it is exposed to during

its passage through the pelvis; they consist in
the shortening of certain diameters of the head,
and in the elongation of others. In vertex presentations the occipito-frontal, sub-occipito-bregmatic, bi-temporal, and bi-parietal are diminished;
while the maximum diameter of Budin, or the line
running from the chin to a point on the sagittal
suture midway between the apex of the occipital
bone and the large fontanelle, is increased. These
changes are enabled to occur by the presence of
sutures and fontanelles. . One parietal bone is
depressed and the other overlaps it. The occipital
bone slides under the two parietal bones, and the
frontal bone does the same. The cartilage, between
the squamous and the petrous portions of the temporal
bones, allows the former to be pressed inwards, and
thus acts as a hinge.

The caput succedaneum is a tumour due to a serous
infiltration of the connective tissue, which occurs
over the presenting part. The latter is the only part of
the child which is not subjected to pressure; and,
accordingly, there is a transudation of serum into its
connective tissue. A small quantity of extravasated
blood is also present in the tumour, due to the
rupture of minute vessels.

Mechanical Phenomena.—The mechanical phenomena include,—the series of passive movements which
the fœtus performs, in order to adapt itself to the
varying curves and diameters of the genital canal,
and the various changes in position which are forced
upon the bones of the pelvis by the passage of the
fœtal head through it. As each presentation has
its own mechanism they will be described together.

Accordingly, here, under the head of normal labour, comes the mechanism of a vertex presentation.

The Movements of the Pelvic Bones.—I shall first consider the movements of the pelvic bones. As mentioned in a previous chapter (*v.* p. 7), the pelvis, although at other times a rigid structure, permits during parturition certain movements to take place at its joints. At the symphysis pubis the fibres composing the interpubic ligaments soften and elongate, so that a very slight degree of separation is allowed. At the sacro-iliac joints a certain amount of mobility is permitted, so that the sacrum can rotate about these joints on an antero-posterior plane. As the head enters the brim, the promontory is pushed backwards, and thus the conjugate diameter is increased. As the head descends into the pelvis, the promontory returns to its original position, and then becomes depressed, owing to the rotation backwards of the lower pieces of the sacrum under the pressure of the head at the outlet. The sacro-coccygeal joint permits the greatest range of movement. It is a hinge joint, so arranged that the coccyx can be pressed backwards by the head, thus increasing the antero-posterior diameter of the outlet by about three-quarters of an inch. In some cases this movement may take place at one of the intercoccygeal joints instead of through the sacro-coccygeal joint.

The Movements of the Fœtus.—The movements by which the fœtus is adapted to the curves of the genital canal can be resolved into five distinct groups :—

  (1) Descent ; and, coincidently,

  (2) Flexion.

(3) Internal rotation.

(4) Extension.

(5) External rotation.

(1) *Descent.*—As the uterus contracts, the presenting head is driven down into the brim of the pelvis. It enters the latter in such a manner, that, its bi-parietal diameter is parallel to one or other of the oblique diameters of the pelvis, according to the position in which the child lies. In describing the mechanism of labour here, I shall suppose that the foetus lies in the first position. In that position the head enters the brim with its bi-parietal diameter parallel to the left oblique diameter of the pelvis. When the head first comes to the brim, it is in a position of unstable equilibrium. As it is driven down by the contractions of the uterus, it is obliged to assume a position of stable equilibrium—that is, it must either flex or extend.

(2) Of the two alternatives which I have mentioned, *flexion* occurs in more than 99 per cent. of all cases of head presentation. Why should it occur so very much more frequently than extension? Let us imagine the antero-posterior diameter of the head to be a rod A which is fixed by a moveable joint C on to another rod B, *i. e.* the spinal column (*v.* Fig. 4). Now, if the joint C be situated in the centre of A, equal resistance offered to the ends of A as it is driven downwards, will not cause it to change its position as regards B. But if the joint C be nearer one end of A than the other, then equal resistance offered to the ends of A will cause the long arm to approach B, the short arm to rotate in the opposite direction. And this is practically what occurs to the

head as it enters the brim. The antero-posterior diameter of the head is pivoted on the spine in such a manner that its anterior arm is the longer. The resistance of the brim is equal on both occiput and forehead, with the result that the head moves so as to cause the chin to approximate the chest. In this way flexion commences, and continues until the head comes into a position of stable equilibrium, with the chin resting upon the chest. There are other ex-

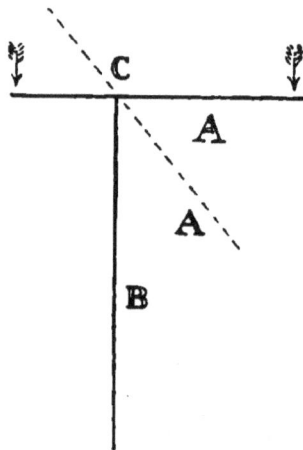

FIG. 4.—Diagram showing how flexion is produced.

planations of the occurrence of flexion, but this is the theory usually accepted. The result of flexion is that a shorter diameter of the head is substituted for a longer one. When the head is flexed the longest diameter which has to pass through the brim is the sub-occipito-bregmatic. Before flexion occurred the presenting diameter was the occipito-frontal, and so a diameter of 3¾ inches has been substituted for one of 4½ inches.

(3) *Internal Rotation.*—When flexion has occurred,

the head is advancing with its vertex presenting, and its sub-occipito-bregmatic diameter lying in the right oblique diameter of the pelvis. It continues in this diameter until the presenting part reaches the pelvic floor; when the occurrence of internal rotation of the head brings the sub-occipito-bregmatic diameter to lie in the conjugate diameter of the pelvis. This internal rotation is brought about by two factors. The first is the inclination of the pelvic floor. I have already described the anterior inclined planes of the pelvis; they slope downwards and forwards, so that whatever impinges first upon them is guided to the front. The second factor is, that, the point of least resistance to the advance of the vertex lies in front. Posteriorly there is the resistance offered by the firm vaginal walls and perinæum; anteriorly there is the arch of the pubes, under which there is a minimum of resistance. The fact that it is always the lowest portion of the head which rotates forward, favours both these theories. Accordingly the agents which tend to cause forward rotation of the occiput are :—

(1) Flexion, which brings the occiput lowest.

(2) Good labour pains, which drive the presenting part onward.

(3) A firm perinæum, which, by causing a maximum of resistance posteriorly, compels the occiput to turn in the direction of least resistance (Lusk).

Extension.—When internal rotation is completed the head is lying so that its sub-occipito-bregmatic diameter approximately corresponds to the conjugate diameter of the pelvis; and the occipital bone is under the pubic arch. Now the advancing head

has to travel in a different direction from that in
which it started, in order to suit itself to the forward
curve of the genital canal.   During the movements
of internal rotation, the head has been commencing
to adapt itself to this curve, and now it advances
along it, and at the same time delivers itself, by the
movement of extension.   In this stage the occiput
of the child becomes fixed beneath the pubic arch,
and the head, as it extends, rotates round this fixed
point, in such a manner that the chin leaves the chest,
and the face slowly appears from behind the
perinæum.   The factors in the extension of the head
are two :—first, the uterine contractions, which force
the head to move in the direction of least resistance,
in accordance with the law given above; and
secondly, the contractions of the levator ani muscle.
This muscle is so situated that it forms part of
the floor of the pelvis, and of the posterior wall
of the vagina.   When uncontracted its anterior
surface is concave; when contracted it tends to
become flat, and so to push forward anything which
may be lying upon it.   Just previous to extension,
the forehead of the child lies in the concavity of
this muscle; accordingly, as the muscle contracts it
forces the forehead forwards.   Thus extension of
the head is brought about, and by the time it is
completed, the chin has appeared from behind the
perinæum, i. e. the head is born.

External Rotation.—This is the final movement of
the head.   It consists of two parts—(1) restitution,
(2) external rotation.   When the movement of internal rotation occurs, the head rotates to suit the
pelvis, but the shoulders do not take part in this

movement. Thus the head becomes slightly twisted as regards the shoulders. This position continues so long as the head is subjected to the pressure of the pelvis. As soon as the head is freed from this pressure, its first movement is to rotate, so as to lie in its normal relationship to the shoulders, *i. e.* restitution occurs. As the head travelled through the pelvis, the shoulders became engaged in the brim in the diameter at right angles to that in which the head engaged. Thus, in a first position, the shoulders engage in the left oblique diameter of the pelvis. As they descend, the anterior shoulder, being slightly lower than the posterior one, rotates in front, so as to lie in the conjugate diameter of the pelvis. It is this movement that causes the completion of external rotation, the already delivered head rotating to suit the new position of the shoulders. Usually the head rotates in such a manner as to return to its former position ; *i. e.* in a first position, it rotates with the occiput pointing to the mother's left thigh ; in a second position, with the occiput pointing towards the right thigh.

When the anterior shoulder has rotated in front, it becomes fixed under the pubic arch. The posterior shoulder then sweeps over the perinæum, and is born. The arms follow, folded upon the chest ; and the rest of the body, being smaller than that which has gone before, is born without further difficulty. It is interesting to note that these different movements are, so to speak, complementary to one another. Thus first occurs flexion of the head ; then internal rotation ; then extension, the complement of flexion ; then external rotation, the complement of internal rotation.

5

CHAPTER VIII.

MANAGEMENT OF NORMAL LABOUR.

Diagnosis of Vertex Presentation—Treatment—Methods of pre-
serving the Perinæum—Care of the Funis after the Birth of
the Head—Management of the Third Stage—Ligation of the
Cord—Expulsion of the Placenta—"Credé's Method"—Method
of determining whether the Placenta be in the Uterus or in
the Vagina—Use and Abuse of Ergot—Abnormal Mechanism
in Vertex Presentations.

Diagnosis.—The diagnosis of vertex presentations
is best made by *abdominal palpation.* By this means
the head is discovered to be lying in the lower
uterine segment, and either just above the pelvic
brim or engaged in it. The chin lies at a higher
level in the uterus than the occiput; thus denoting
that the case is one of vertex presentation, and
not of face or brow (*v.* Fig. 3). The breech is at the
fundus, and between it and the head lies the body,
usually inclined to one or other side. The limbs may
or may not be felt, according as the back is posterior
or anterior. By *auscultation,* the point of maximum intensity of the heart-sounds is found to be
below the umbilicus, and to one or other side of the
middle line, according to the side at which the
back lies. By *vaginal examination,* a round tumour
is found to be presenting, and upon it the sutures
and fontanelles can be felt. The anterior fontanelle

is recognised by its lozenge-like shape; the posterior is smaller and triangular. If, however, the bones overlap one another, owing to moulding, the fontanelles may be obliterated. Their site can then be re-cognised by the fact, that, a number of sutures meet at a point. At the anterior fontanelle, four sutures meet; at the posterior, three.

Treatment.—The treatment of the three stages must be considered separately.

First Stage.—The first stage commences with the onset of labour pains, and ends with the full dilatation of the os. Its chief physiological pheno-menon is the occurrence of intermittent contractions of the uterus, which tend to drive the ovum into, and so to dilate, the cervical canal. The treatment consists in keeping up the patient's strength, in helping nature in a natural way, and in avoiding meddlesome and dangerous interference.

Concerning the first of these it is unnecessary to say much. The patient must get suitable stimu-lating food at short intervals, and anything likely to derange the stomach must be avoided. At the commencement of labour, when the pains are few and far between, she should have some occupation which will keep her mind off her condition, and so prevent useless fretting.

The second indication—to help nature in a natural way—is easily carried out. We can help nature to dilate the cervix, by keeping the woman in such a position that the action of gravity aids the pains in driving the ovum downwards against the os; in other words, by making the patient walk about or sit in a chair, and not allowing her to lie down unless she feel tired. Indeed, nature itself will prompt her to maintain an upright posture. Moreover, the uterus must be in such a position, that its contractions can act to the greatest advantage. If the abdomen be pendulous, or if any degree of lateral obliquity of the uterus be present, the contractions will drive the head, not into the pelvic cavity, but against the brim. This obliquity, or anteversion, of the uterus is best corrected by an abdominal belt or binder, tightly applied, so as to keep the uterus in a proper position. Another important point is to remove all obstruction to the descent of the head. In a normal case the only obstructions present are a full bladder, or a loaded rectum. To avoid the former, the patient must be made to pass water frequently; or, if necessary, a catheter must be passed. To insure that the rectum be empty during the second stage, a purgative should be given as soon as the first symptoms of labour appear, followed in a few hours by a soap-and-water enema. It is well to repeat the latter as soon as the patient gets into the second stage, to avoid the soiling caused by fæces being forced out by the descending head. It is useless to endeavour to get the patient to "bear down," i. e. to voluntarily contract her abdominal muscles during the first stage. As soon as voluntary efforts will have any effect, that is, as soon as the os is dilated, she will "bear down" of her own accord. Premature efforts only waste her strength, and make no impression upon the cervix; inasmuch, as they tend to drive the entire uterus and its contents into the pelvis, and not to force the ovum against the cervix. Indeed, by the absence, or presence, of voluntary bearing-down efforts, we can tell whether the patient be in the first or second stage, without making any vaginal examination.

The third indication is to avoid meddlesome and dangerous interference. This includes superfluous vaginal examinations; manual or instrumental dilatation of the os; the application of forceps, when they are not only unnecessary, but absolutely contraindicated; and prophylactic vaginal douching, when it is not required (v. page 4). I have already spoken of the advantages of abdominal palpation over vaginal examination, and of the dangers of the latter. One vaginal examination, however, is necessary, in order to determine if the cord be presenting or prolapsed. It should be made, if possible, about the end of the first stage; and then, if the os be fully dilated, the membranes may be ruptured at the same time.

Second Stage.—The second stage commences with the full dilatation of the os, and ends with the birth of the child. Its chief physiological phenomena are the continuance of the involuntary and intermittent uterine contractions, with the added help of voluntary contractions of the abdominal muscles, the diaphragm, and, indeed, of most of the muscles of the body. The result of these contractions is:—first, that the membranes rupture, having lost the support of the cervix; and secondly, that the foetus advances downwards through the vagina, presses on and dilates the perinæum, and finally is born. The indications for treatment are the same as before, until the head appears at the vulva. As, however, the physiological phenomena have changed, so the manner of carrying out the indications changes also; and we endeavour to help nature in a different way from the method adopted in the first stage. As the os is fully dilated, and as voluntary " bearing-down " efforts are now occurring, we must put the patient in such a position that she can make the most use of her strength. This she can best do in bed. Let the patient lie on her side, with her feet against the end of the bed, and give her something on which she can pull. A towel tied to the end of the bed is best; as, by pulling on it, she can counterbalance the force with which she is straining against the end of the bed. At the same time instruct her to hold her breath during a pain, and to " bear down " with all her strength.

As soon as the head appears at the vulva, the treatment becomes more active; and the obstetrician prepares to deliver the patient. Now, the chief indication is to avoid rupture of the perinæum, and we must place the patient in the best position for attaining this end. There are two positions from which to choose :—the patient may lie either upon her left side or upon her back. Of these the first is by far the more preferable. It is cleaner, the progress of the head can be more satisfactorily watched, and more effectual measures can be taken for the preservation of the perinæum. Having the patient, then, in the side position, what method shall we adopt to prevent the occurrence of perinæal laceration ? Numerous methods have been proposed ; they may be divided into direct and indirect. The direct method consists in directly supporting the perinæum with the hand, with the object of preventing it from becoming over-distended, and so lacerated. It consists in laying the palm of the hand on the perinæum, with the concavity between the first finger and thumb directed so as to enclose the posterior end of the vulva, and then pressing the perinæum upwards against the advancing head. It is a bad method, and opposed to the natural process by which the perinæum is made dilatable. That is, it prevents the serous transudation from being poured into the connective tissues of the perinæum, by compressing the latter between the hand and the advancing head. The indirect method consists in endeavouring to push the head forward as much as possible. This can be done either by introducing two fingers into the rectum, or, better still, by applying the hand behind the anus, and pushing the head forward. I shall describe the latter method in full, as I consider it the best of all methods for the preservation of the perinæum. To understand it fully it is a help to study the accompanying diagram (v. Fig. 5).

The rod A B represents the fœtal body which is being driven downwards by the uterine contractions in the direction shown by the arrow C. This direction causes the head to press upon the perinæum P. If the end A of the rod can be pushed forwards towards A' then the uterine contractions will drive the rod in the direction shown by the arrow C', that is through the vulva. Accordingly, any pressure applied in the direction of the arrow H will take a proportionate amount of pressure off the perinæum. One point must be remembered;—the parturient canal is in the shape of a curve, with the concavity

forwards. This curve may be considered as consisting of an upper segment and a lower segment.
While the fœtus is advancing in the upper segment
of the curve it is being driven in the direction of a
point midway between the anus and the tip of the
coccyx. As it comes into the lower segment it
changes its direction, and moves towards the vulva.
If forward pressure be applied to the advancing
head, while it be still in the upper segment of the

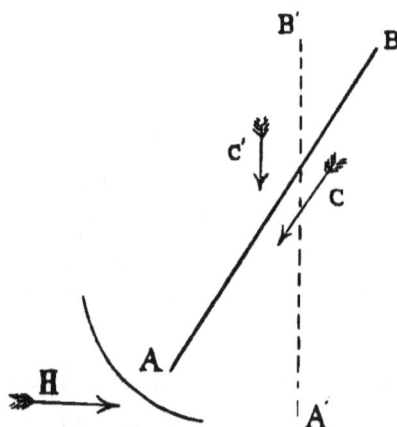

FIG. 5.

curve, such pressure will drive it back into the
uterus. If, however, we wait until the head gets
into the lower segment, then, our forward pressure
will push the head off the perinæum and in the
direction of the vulva. Accordingly this pressure
can only be of use when the head has passed the
" sticking point," if I may call it so,—that is the point
of junction of the two portions of the parturient curve.
This is the theory of the method ; it is carried out
as follows :—The patient is in the side position, with

her buttocks well over the edge of the bed. The
physician stands by its side facing its foot, and
passes his left hand over the patient's hips, and then
between the thighs from in front. With this hand
he endeavours to draw forward the advancing head,—
not, as some advise, by levering it out from behind
the perinæum, but by applying the fingers to the
scalp alone and drawing it forwards. Of course
this cannot be done, effectually, until the head is
sufficiently advanced to be able to get some purchase
upon it with the fingers. Meanwhile, the right
hand is idle, waiting until the head is sufficiently
low for forward pressure to be of avail. Then the
heel of the hand is applied between the anus and
the tip of the coccyx, and the head pushed forward
and delivered at a suitable moment. There are two
essential details, the due observance of which tends
greatly to aid the preservation of the perinæum.
First, the head must not be allowed to extend too
soon. Extension should be delayed until the lowest
possible point of the occiput comes to lie under the
symphysis; as the nearer to the neck the point of
the occiput is, round which the head rotates, the
smaller will be the sagittal diameter of the head,
that will distend the perinæum. This is carried out
by pressing the forehead and face forward, in such
a manner as to keep the chin, at the same time, in
contact with the chest. The exact method of doing
this, as well as the knowledge that the head is past
the "sticking point," can only be attained by
experience. The second essential is to deliver the
head between the pains, and not during a pain.
This is done by trying to check the voluntary efforts

of the woman at expulsion; by making her cry out
instead of holding her breath; and by taking away
the support for her feet, and any towel or rope upon
which she may be pulling. Then, when the pain
has passed off, the head may be pressed out as
already mentioned.

The moment the head is born, we must determine
whether the cord be twisted round the neck or not.
This is done by passing a finger or two into the
vagina, and feeling carefully round the neck and
upwards as far as the shoulder. If the cord be
there, it is readily felt, and must be immediately set
free. The danger of leaving it is, that, if it be a
short cord, or if it be several times round the neck, it
may not be sufficiently long to permit of the birth
of the child without the tearing of the cord. It
can be set free in any of three ways. The usual
and easiest method is to pull down a loop of it, and
pass this loop over the head of the child; if there
be a second turn round the neck, it must be pulled
down and set free in the same manner. If the
cord be drawn so tightly round the neck, that it
cannot be slipped over the head, we try to slip it
over the shoulders. To do this, the nurse presses
upon the fundus, and so drives the child downwards;
as it advances the cord is slipped, first, over one
shoulder, and then over the other, and so the child
is expelled through the loop of the cord. If
neither of these methods can be performed, owing
to the excessive tightness of the cord, divide it
with scissors; and deliver the child instantly, by
means of pressure upon the fundus, and traction
upon the head. While the cord is being set free,

the nurse should wipe carefully the eyes of the child, to remove any discharge that may have got into them during the passage of the head through the vagina. This is an important prophylactic measure in the treatment of the purulent ophthalmia of infants. When the cord has been set free, the remainder of the delivery may be left to nature, provided that the cord be still pulsating; if not, the child must be instantly delivered. In accomplishing this avoid undue or premature traction on the head, as it may hinder rotation. *Vis a tergo* is much to be preferred to *vis a fronte*. That is, press upon the fundus, and, as the shoulders come down, lift the child upwards towards the mother's abdomen, so as to allow the posterior shoulder to sweep over the perinæum. Then depress the body again slightly, in order to bring the anterior shoulder from beneath the symphysis. Once the shoulders are born the rest of the child quickly follows, as it is smaller than the part which has gone before.

Third Stage.—The third stage commences as soon as the child is born, and ends with the birth of the placenta. The physiological phenomena are, still, the occurrence of intermittent contractions, and of permanent retraction, of the muscle-fibres. The result of these is, that the placenta is detached and expelled from the uterus, while, at the same time, the open mouths of the vessels are closed, and hæmorrhage thus prevented. Before studying the treatment of the third stage, we must first dispose of the newly born infant. In a normal case it will begin to cry as soon as it is born ; if not, any slight cutaneous stimulation will cause it to do so. A dash of cold water or a couple of smart slaps of the hand, are the time-honoured methods. If there be mucus in the throat, it must be cleared out *before* attempting to make the child inspire. Lastly, the cord has to be tied, and the child thus separated from the mother.

Formerly, it was a subject of great dispute, whether the cord should be tied the moment the child is born, or whether the application of the ligatures should be deferred until the cord has ceased to pulsate ;—that is, early *versus* late ligation of the cord. According to Professor Budin's experiments, the child receives an additional four ounces of blood by adopting late ligation ;—that is, early ligation deprives it of a corresponding amount. It is obvious ; that the child receives this blood from the placental vessels. But how does it receive it, and is it to its advantage that it should do so ? It may obtain the extra blood in two ways. Either the contractions of the uterus compressing the placenta may drive the blood, from the latter, into the child's vessels ; or, the first inspiration of the child, by opening up the pulmonary circulation, may create a negative pressure in the great vessels near the heart, and so may cause blood to be sucked in from the placenta. As Professor Lusk points out, it is of practical importance which of these alternatives is the correct one. If the blood be forced in by the contractions of the uterus, the vessels may become over-distended, and a distinct danger to the child be created. If, on the other hand, the influx of blood be due to thoracic aspiration, then the placental blood is going to supply a distinct want on the part of the child ; and the loss of it is absolutely injurious. The result of experimental research is to prove, that the latter alternative is the more likely ; and, furthermore, children in whose case late ligation of the cord has been adopted, are stronger and healthier than those in whose case it has been ligatured the moment they were born. The cord, therefore, should not be ligatured until it has ceased to pulsate. It is then tied with a double ligature, one applied two inches from the umbilicus of the child, and the other as close as possible to the vulva. Before applying the second ligature, first draw gently on the cord, so as to pull out any loops that may be lying in the vagina ; the object of this will be explained presently. The cord is then divided half an inch above the first ligature, and the child taken away.

The indication for treatment in the third stage is to promote contractions of the uterus ; in order to cause expulsion of the placenta, and of any clots that may be present, and to prevent hæmorrhage, or the admission of air, into the uterine cavity. With these objects in view, we turn the patient on her back the moment the child is born, and grasp the fundus of the uterus with the hand. This hand must be so applied as to grasp the entire fundus. It thus controls the uterus in the intervals between the contractions, and prevents it from filling with blood. If only the anterior surface of the uterus, or the lower segment, be seized by the hand, we are apt to stimulate this portion of the uterus to contract, and thus cause irregular contractions, or the so-called hour-glass contraction. A good method of applying the hand is to sink it with its ulnar border downwards into the abdomen, until it touches the promontory of the sacrum.

Then the entire fundus lies in the concavity of the hand, and it is impossible for it to escape upwards. Having the uterus under control, the placenta must next be considered. Two questions may be asked with regard to its removal. First, how is it to be removed ?   Secondly, when is it to be removed ?

How is the placenta to be removed ?   It can be removed by pulling upon the cord, and thus dragging it out of the uterus ; by passing the hand into the uterus and taking it away ; by the so-called "Credé's method"[1] of expression from above; or by leaving its expulsion to nature. I shall consider the last method first.   Let us see how nature delivers the placenta.

A few moments after the birth of the child the uterine contractions return.   The uterus as a result becomes much smaller, and the placental site becomes very greatly reduced in size, so that it is no longer large enough for the placenta.   The latter, being too dense to be crumpled up, so as to suit its reduced area of attachment, becomes detached, and lies loose in the uterus.   Then the next pain, or perhaps the same pain that detached it, causes it to be expelled into the vagina.   It lies there for some time, and is gradually worked downwards, helped by any contractions of the abdominal muscles that may occur.   This is a tedious process, and lasts on an average two or three hours.

[1] The delivery of the placenta by pressure upon the fundus, as distinct from traction on the cord and from manual removal, I here allude to as "Credé's method," because by this title it is most generally known. In reality it was practised in the Rotunda Hospital, and described by M'Clintock and Hardy in their 'Practical Observations on Midwifery' (1848), many years before Credé taught its use.

Expression by "Credé's method," before the placenta is detached, will materially shorten the third stage. It is, however, very liable to cause post-partum hæmorrhage, as the uterine fibres will not have had time to retract properly, and so to close the mouths of the vessels; also small portions of placenta are frequently left behind. If, however, expression be delayed until the uterus of itself has expelled the placenta into the vagina, then " Credé's method " of expression is a most important mode of treatment, and a perfectly safe one.

There are the same objections to manual removal by introducing the hand into the uterus, if done before the placenta is detached, as to expression; with the added objection that the risk of sepsis is very much increased. When the placenta is in the vagina, manual removal has no advantage over expression, but many disadvantages.

Traction on the cord is the worst of all methods of removing a placenta which is still adherent. As the cord is inserted into the centre of the placenta, traction causes detachment of the latter, at first, in the centre. A cavity is thus formed behind the placenta, into which blood is sucked as the cord is pulled upon. If the adhesions between the uterus and placenta be so dense as to prevent separation, strong traction on the cord may cause inversion of the uterus. If the placenta have left the uterus, then traction on the cord may be employed to complete delivery. It has, however, no advantages over expression, and is more likely to favour the retention of portions of the membranes.

The safest and best treatment, then, is to leave

the first part of the expulsion of the placenta to
nature—that is, its detachment and its expulsion out
of the uterus into the vagina.   As soon as it gets
there, its further progress can be expedited most
safely by adopting expression, as in "Credé's method."

The foregoing also answers the second question,
when is the placenta to be expelled ?   The placenta
is to be expelled as soon as it has left the uterus.
Premature expulsion exposes the patient to the

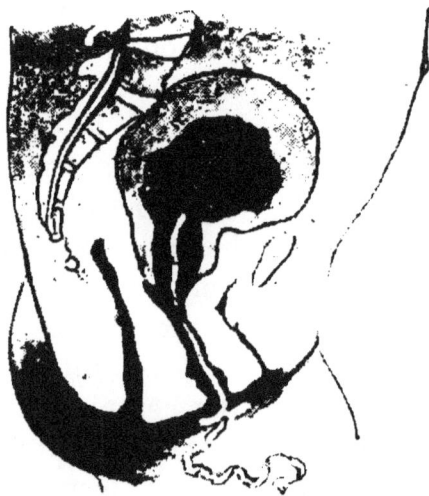

FIG. 6.—Semi-diagrammatic representation of the condition of the
parts before the expulsion of the placenta from the uterus.

danger of post-partum hæmorrhage, and subsequent
sapræmic trouble.   Leaving the entire process to
nature, means keeping the patient from the rest she
requires, for a much longer period than is necessary.

How is it possible to tell when the placenta has
left the uterus ?   There are four indications :—

(1) *The cord lengthens.*   When the cord is being

tied two ligatures are used,—one close to the umbilicus
of the child, the other as close as possible to the
vulva of the mother, having first pulled upon the cord
slightly, to withdraw any portion of it which may be
coiled up in the vagina (*v.* Fig. 6). As the placenta
leaves the uterus and comes into the vagina, it is
obvious that the portion of cord outside the vulva will

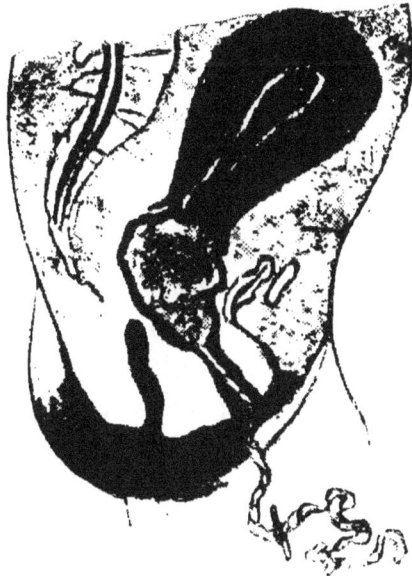

Fig. 7.—Semi-diagrammatic representation of the condition of the
parts after the placenta has been expelled into the vagina.

be increased in length ; and so the ligature, which
originally was tied as close to the vulva as possible,
will come to lie four to six inches away from it
(*v.* Fig. 7).

(2) *The fundus rises upwards to about the level of
the umbilicus.* When the child is born the uterus
sinks into the pelvis, under the pressure of the

abdominal muscles, and the controlling hand of the assistant.   It thus telescopes itself into the vagina, as is shown in the diagram (v. Fig. 6).   As the placenta is expelled into the vagina, it comes to occupy the place where the uterus was; and, consequently, the latter is pushed upwards out of the vagina, and the fundus is felt almost at the level of the umbilicus (v. Fig. 7).

(3) *The abdominal wall bulges forward, above the pubes.*   This is due to the presence of the placenta in the vagina, or in the lower uterine segment.   The placenta lying there pushes forward the anterior vaginal wall, and in front of it the bladder and the abdominal wall, and thus causes an appearance resembling a full bladder.

(4) *The mobility of the uterus is increased.*   When the uterus is lying in the pelvic cavity with the placenta inside it, it is supported on all sides by the pelvic brim, and cannot readily be moved from side to side (v. Fig. 6).   But, as it rises, it loses this support, and becomes balanced—if I may use the term—on the top of the placenta, and so can be moved about with ease from side to side.   This is well shown in the diagram (v. Fig. 7).

As soon as we know by these signs that the uterus is empty, the placenta may be expressed by " Credé's method."   To do this, grasp the fundus with one or both hands during a pain, and press it downwards and backwards, in the direction of the second piece of the sacrum.   By this means, the uterus is pressed down into the vagina, and the placenta driven out. The latter is immediately supported by the hands of the nurse, and twisted round several times so as to cause the membranes to become detached; they are thus brought away entire.

Let me repeat the management of the third stage in a few words. The moment the child is born the patient is turned on her back, and the doctor or nurse controls the fundus with one hand. As soon as the cord has ceased pulsating, it is ligatured as described above, and the child is removed. Nothing further is done until the placenta has left the uterus. As soon as this occurs, the placenta is expressed from the vagina, seized in the hands, and twisted round so as to bring away the membranes entire. The third stage is thus completed, and nothing remains but to wash all blood off the patient, to apply a diaper wrung out of corrosive sublimate to the vulva, and a tight abdominal binder. During all this time the hand must be kept on the fundus, in order to prevent it from becoming full of clots. It should not be removed until the last pin of the binder is inserted.

There has been much controversy regarding the exact manner in which the placenta is delivered. Schultze maintains, that, as the placenta is detached, a hæmatoma is formed behind it, and that this hæmatoma forces the placenta downwards into the membranes, which are thus inverted. The placenta thus comes out with its fœtal surface outside, and the membranes are turned inside out. Matthews Duncan, on the other hand, said that this method never happens unless the cord be pulled upon. He thought that the placenta comes out edgewise, in the same manner as a button goes through a button-hole. It is not a matter of great practical importance as to which is the mechanism of delivery. As a matter of fact Schultze's method occurs in about three-quarters of all cases ; but then there is frequently some traction upon the cord, as the child is born.

Ergot.—The knowledge of the use of ergot is of great importance. It is undoubtedly of service at certain times, but given at the wrong time it is most dangerous. Its action is to cause tonic contraction of the entire uterus. Considerable importance must be attached to the fact, that the contractions due to the action of ergot are tonic and not intermittent ; and that, as such, they differ from the physiological contractions of the uterus. This fact indicates the time at which ergot may be given. It may be given, when tonic contraction of the uterus is not dangerous to mother or child ; and, with a few exceptions, this is only when the uterus is empty. If ergot be given during the first stage of labour, it will delay the dilatation of the os, cause dangerous pressure upon the child, and even detachment of the placenta. In the second stage ergot is dangerous, unless it be absolutely certain that there is no obstacle to the rapid birth of the child. This excludes its use in most cases. In the third stage it is also contra-indicated, as it may cause incarceration of the placenta. Then, if hæmorrhage occur, the condition of affairs is very serious. Hæmorrhage, occurring during the third stage, usually requires the removal of the placenta; but, if ergot have been given, the cervix will be so tightly contracted that it may be impossible to do so. When, therefore, may ergot be used ? It may be used when the uterus is empty, in order to promote tonic contraction ; and it is in very few cases that it is advisable to use it during any of the three stages of labour. It may be given as a routine treatment as soon as the placenta comes into the vagina, or even a little sooner, if we be prepared to take away the placenta before the ergot commences to act. Ergot, given by the mouth, causes uterine contractions in from ten to fifteen minutes. Ergot, given hypodermically, acts in five minutes, or even less.

Abnormal Mechanism in Vertex Presentations.— In some cases of vertex presentation, during the occurrence of internal rotation, the forehead, and not the occiput, rotates in front under the symphysis. This condition is probably due to incomplete flexion on the part of the head, which causes the forehead to lie at a lower level than the occiput. Then, following the rule that the part of the child, which first comes on to the pelvic floor, rotates in front, the forward rotation of the forehead occurs. The result of this condition is, that labour is very much more tedious than is usual, and frequently has to be terminated instrumentally (*v.* page 174). During the birth of the head the perinæum is more distended than usual, and laceration frequently occurs.

CHAPTER IX.

NATURAL PRESENTATIONS.

Pelvic Presentations : Proportion — Positions — Mechanism — Diagnosis—Result of the Head not filling the Lower Uterine Segment—Ætiology—Treatment—Method of bringing down an Extended Arm—Methods of delivering the After-coming Head; the Prague Method, Martin's Method, the Smellie-Veit Method—Abnormal Mechanism in Breech Cases—Face Presentations : Proportion—Positions—Ætiology—Mechanism — Diagnosis — Treatment — Abnormal Mechanism in Face Presentations.

NATURAL presentations are those presentations which can deliver themselves without the presentation changing, with the exception of vertex presentation which is known as normal presentation. Under the term natural presentation are included pelvic and face presentations.

PELVIC PRESENTATIONS.

Pelvic presentations include all cases in which the lower pole of the fœtus presents. Most frequently it is the breech which is lowest, but sometimes it may be the foot or the knee. The proportion of cases in which breech presentations occur varies from 1 in 80, in primiparæ, to 1 in 23, in multiparæ; knee presentations occur about once in 800 births.

In considering the mechanism no difference need be made for knee and foot presentations.

Positions.—Two positions are recognised for breech presentations : *first,* when the back of the child is on the left : *second,* when the back is on the right.

Mechanism.—The dimensions of the breech are not of any very great importance. They are considerably smaller than the dimensions of the head, and can be reduced still further by compression. The bi-trochanteric diameter is the greatest, and measures 3½ inches. The sacro-pubic diameter measures 2 inches. The mechanism of the breech is very simple. It engages with the bi-trochanteric diameter in the oblique diameter of the pelvis. As it descends, the anterior hip usually lies at a slightly lower level than the posterior. The former thus reaches the pelvic floor first, and, as a result, rotates in front and lies under the symphysis. The posterior hip rotates round it, sweeps over the perinæum, and is born. The rest of the trunk then follows in the same manner. The attitude of the child is the same as in a vertex presentation; consequently the feet generally come out close to the breech, and the arms are folded across the chest. The head enters the brim with its sub-occipito-bregmatic diameter lying in the oblique diameter of the pelvis, and the chin flexed upon the chest; thus the shortest diameters of the head engage. As it descends, the occiput rotates in front, the chin being still kept closely applied to the chest, owing to the pressure of the coccyx and perinæum. The occiput now rests under the pubes,

while the face rolls out over the perinæum, the
chin appearing first, then the mouth, nose, eyes,
forehead, and occiput.

Diagnosis.—The diagnosis can be made by
abdominal palpation and by vaginal examination.

By *abdominal palpation* we determine that the
fœtus is presenting by one of its two poles, and
that the opposite pole is at the fundus. The pole

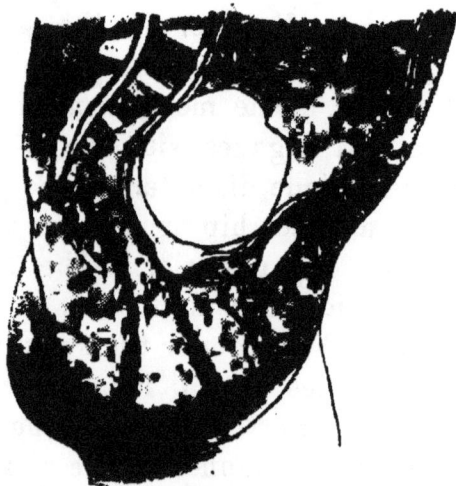

FIG. 8.—Result of the head filling the lower uterine segment exactly
—semi-diagrammatic (modified from 'The Norris Text-Book of
Obstetrics').

at the fundus is round and hard, and ballotts
easily from side to side independently of the back ;
there is, moreover, a groove between it and the
back. This distinguishes it at once as the head.
To confirm this diagnosis we palpate the presenting
pole, which is also round and hard, but which does
not ballott nor move independently of the back ;

and, in a favourable case, the legs can be felt springing from it.

By *vaginal examination* the diagnosis can also be made, when labour is a little advanced, but it is rather more difficult to do so. At the commencement of labour, owing to the tardy fixation of the breech, the presenting part cannot be reached with the finger. At this stage the point most likely to attract attention is the peculiar way in which the membranes bulge.

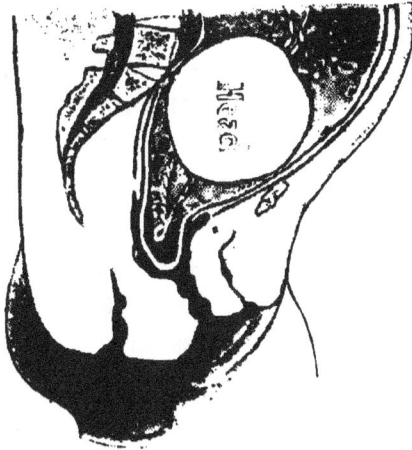

FIG. 9.—Result of the head not filling the lower uterine segment exactly, thus permitting undue pressure on the membranes, and favouring prolapse of the cord—semi-diagrammatic (modified from 'The Norris Text-Book of Obstetrics').

During a contraction they project downwards into the vagina as a cone-shaped tumour. This undue protrusion of the membranes is never seen in a normal case, but always can be noticed when the presenting part or the pelvis is abnormal. In a normal case, the presenting head fills the lower uterine segment completely; consequently, before rupture of the membranes, the liquor amnii in front of the head is completely shut off from the liquor amnii which surrounds the body. When a contraction occurs, the head acts as a ball-valve, and prevents any more liquor amnii from being driven down in front of it (v. Fig. 8). Consequently, the tension on the membranes is increased only in proportion as the head advances, and their premature rupture is avoided. If, however, the presenting part do not fill the lower uterine segment exactly, owing to its irregular shape,—as in face, breech, or transverse presentations,—or if it be prevented from descending into the lower uterine segment,—as in the case of contracted pelvis,—then there is free communication between the waters in front of the presenting part and the waters behind it. The result of this is, that, when a contraction occurs, the liquor amnii round the body is driven downwards in front of the presenting part, and the pressure on the membranes is very greatly increased. This increased pressure causes at first undue bulging of the membranes downward, and, when the os has become partly dilated, premature rupture of them (v. Fig. 9). When, therefore, this cone-shaped projection of the membranes is present, we immediately suspect that there is something abnormal, either in the pelvis, or in the presentation.

When labour is more advanced, the presenting part descends within reach of the finger, and can be recognised. It is not at all as easy to distinguish the breech by vaginal examination as is supposed. By it we determine the presence of a large, hard, round tumour, not at all unlike the vertex; it is distinguished from it by the absence of sutures and fontanelles. It may also be mistaken for a face, on which a large caput succedaneum has been formed. The breech may be distinguished by three bony points and by the anus. The bony points are the two tubera ischii and the tip of the coccyx; and they are so arranged as to form the apices of an equilateral triangle. The anus can only be mistaken for the mouth. It is distinguished from the latter by the absence of the alveolar ridges and of the tongue, by the peculiar way in which the sphincter grips the finger, and by the presence of meconium on the finger when withdrawn. If a limb have prolapsed, it may be necessary to distinguish between an elbow and a knee, or a hand and a foot. A knee is easily distinguished from an elbow by its larger size, and by the presence of the tuberosity of the tibia and the patellar ligament. The mobility of the patella is a fallacious sign, as the knee is always flexed when it presents, and so the patella is fixed. The foot can be distinguished from the hand, most easily, by feeling the heel. In default of it, the phalanges are the best guides; in the foot, the line of the tops of the toes is straight; in the hand, the line of the tops of the fingers is curved. Again, the thumb can be apposed and opposed to the palm; the great toe cannot.

By *auscultation* the heart is heard slightly above the level of the umbilicus, and to one or other side of the middle line.

Ætiology.—I have mentioned before the causes of head presentations, namely:—the uterus is of an ovoid shape, and the fœtus in its usual attitude is also of an ovoid shape ; the fundus is the larger pole of the uterus, and the breech is the larger pole of the child. Accordingly, in normal cases the breech is to be found at the fundus, and the head at the pelvic brim. Anything, therefore, which tends to change the shape of the uterus or of the child, may be put down as a cause of mal-presentation, and especially of breech presentations. The principal of these causes are :—

(1) *Multiparous uterus.* The uterine walls have become lax.

(2) *Contracted pelvis.* The uterus is pushed upwards out of the pelvis, and so its shape is altered.

(3) *Twins.* The uterus is over-distended.

(4) *Hydramnios.* The uterus is also over-distended.

(5) *Placenta prævia.* The placenta fills up the lower uterine segment, and so changes the shape of the uterine cavity.

(6) *Hydrocephalic head.* The cephalic pole of the fœtus is larger than the podalic pole.

(7) *Premature children.* The fœtus does not fill the uterine cavity, and, consequently, is not guided into its normal position.

(8) *Tumours, and faulty development, of the uterus.*

Treatment.—Pelvic presentations, if recognised soon enough, can be treated in one of two ways.

(1) They can be turned into a vertex presentation, by external version.

(2) They can be left alone, and delivered as pelvic presentations.

(1) A pelvic presentation is considerably more dangerous for the child than a vertex presentation ;

so that it would appear, at first sight, to be better to turn the child. Before deciding on this, the case must be looked at from another point of view. Breech presentations are generally the result of a definite condition in either the child or the pelvis ; and, in many of these conditions, a breech presentation is more favourable for mother or child, or perhaps for both, than a vertex. Therefore, we must consider whether version is likely to improve the condition of affairs, or to do the opposite. These conditions are,—slight degrees of contracted pelvis and of hydrocephalic head, and placenta prævia. In the two first the after-coming head is more easily delivered than if it came first ; while in the case of placenta prævia, where the proper treatment would be to draw down a leg, we have the leg close at hand. Before turning a breech into a vertex, then, it is well to exclude the presence of any of these conditions. If none of them be present, the child may be turned. The only difficulty in version is to keep the child in its new position. If the turning be done some time before labour begins, the child will turn round again to a breech presentation. The best time to perform version is after labour has started, but before the breech is fixed, or the membranes have ruptured. Then, turn the child by external version (v. page 176) and apply a tight binder to keep it in its place.

(2) If a breech presentation be allowed to persist, the treatment of the case is more difficult. There is a general rule for the treatment, during the first stage of labour, of all abnormal cases. It is to avoid anything that may increase the liability of the mem-

branes to rupture prematurely. I have explained
how it is that premature rupture of the membranes is
the rule in these cases; and, in a breech presentation
especially, the prognosis for the child is worse the
earlier the membranes rupture. The indication then
is to keep the patient quiet during the first stage.
Do not allow her to walk about, but rather keep her
at rest in bed; also avoid vaginal examination, at any
rate during a pain, and do not permit the patient to
bear down until after the membranes have ruptured.
There is no further special treatment required until
the breech appears at the vulva. As the breech
slips from behind the perinæum, the attitude of the
physician is one of "watchful expectancy." There
is nothing to be done in an ordinary case, except, to
lift out the feet as they come, in order that they
may not catch in the perinæum. It is worse than
useless to attempt to express a breech, as we do a
head. Any attempt merely results in pushing it
back into the vagina. Delivery is left to nature
until the child is born as far as the umbilicus; then a
loop of the cord is drawn gently down, and the patient
turned upon her back. There are two reasons for
thus drawing down the cord:—(1) As the body
descends, it compresses the cord against the brim of
the pelvis. This pressure is sufficient to prevent
the cord elongating, proportionately to the descent
of the body. The result of this is, that extreme
tension of the cord occurs, between the umbilicus
of the child, and the portion of the cord which is
caught at the brim. This tension may be so great
as to cause the cord to tear. (2) If we draw
down a loop of the cord and observe its pulsations,

we have an exact indication of the condition of the child.

The patient is now on her back, and everything going on normally, *i. e.* the cord is pulsating. The next uterine contraction drives the child out with the exception of its head, or perhaps may expel it completely. If the head of the child be not expelled by the same pain which expels the shoulders, then assistance must be rendered, as will be shown afterwards. If we can wait sufficiently long to allow the uterus to expel the body of the child, there is little fear of the arms becoming extended above the head. The uterine contractions acting as a *vis a tergo* expel the body, and at the same time keep the arms folded across the chest. But in some cases we cannot wait for the uterine contractions, and are obliged to pull upon the body of the child in order to deliver it more rapidly, *i. e.*, *vis a fronte* is substituted for the natural *vis a tergo*. Then the arms are caught at the pelvic brim, and become extended above the head. The cases in which we cannot wait for the uterus to expel the child are those in which the cord is not pulsating when drawn down. The child is then obviously exposed to the danger of asphyxia, and must be delivered as rapidly as possible. In accomplishing this, the skill and quickness of the accoucheur are tested to the full, and upon them the life of the child depends. Always remember the great difference that exists between the expulsion of the child *vi a tergo*, and the extraction of the child *vi a fronte*. If the arms become extended, the time necessary for the delivery of the case is increased. Never pull upon the body until you have first tried to express the child by pressing upon the fundus, *i. e.* expulsion *vi a tergo*. It is only when this fails, that the body is to be pulled upon. If this course of action be necessary, seize the child by the pelvis and draw it downwards as far as possible, while an assistant at the same time presses upon the fundus. The arms usually become extended, and must be brought down before the head can be delivered.

As the child lies in the oblique diameter of the pelvis, one arm is posterior, and the other anterior. It is always better to bring the posterior arm down first, as there is more room in the hollow of the sacrum for the operator's hand than there is behind the symphysis. To bring the arm down, the body of the child is drawn forwards towards the mother's abdomen, and the entire hand is passed into the vagina, with the palmar surface towards the back of the child. Two fingers are then slipped up along the arm until the elbow is reached. If the forearm be flexed, hook the fingers into the angle of the elbow, and draw it gently downwards over the chest. If the forearm be extended, the fingers must be passed beyond the elbow and hooked over the extensor surface of the forearm ; pressure upon the latter causes it to flex, and so to sweep downwards over the face and chest. The posterior arm is thus delivered ; and next the anterior arm must be brought down, if it be extended. It may be brought down as an anterior arm ; or, which is preferable, the body of the child may be rotated in whichever direction will make the anterior arm become posterior by the shortest route. The second arm is then brought down, in the same manner as the former arm. In some cases one arm may have become so twisted as to lie behind the neck of the child—the nuchal position of the arm. If this happen, the arm may be set free by rotating the body ; or, if this fail, it may have to be fractured before it can be brought down.

The chief danger for the child, in any of these cases, is that the humerus or clavicle may be fractured. The former is most usually broken by attempting to bring the arm down with the fingers upon the middle of the humerus, instead of above the elbow. The clavicle sometimes breaks, when we imagine we are doing everything correctly. This is probably caused by the head of the humerus, as it rotates, being pressed inwards by the pelvic brim, and so tending to approximate the ends of the clavicle.

Now the shoulders are born, and the head alone remains to be delivered. This must be done with the greatest rapidity. It is not only in cases in which traction on the trunk of the child has been made, that the head requires to be delivered artificially. Whenever the head is not expelled by the same contraction which expels the shoulders, it will require assistance. The reason of this is manifest. When the shoulders are born, the head has left the uterus and is lying in the vagina ; accordingly, the uterine contractions have no power to expel it. The head must never be allowed to remain, for a moment longer than is necessary, in the vagina. The cold air, chilling the body of the child, causes premature attempts at inspiration, and mucus and meconium are sucked into the lungs. Further, if the cord have not been compressed up to this point, it is certainly compressed now by the head. And lastly, as the uterus is empty, the placenta is very probably in process of being detached. There are three excellent methods of delivering the after-coming head :—

     (1) The Prague method.

     (2) Martin's method.

     (3) Smellie's method.*

Before describing them one point must be insisted upon. The head must be brought into a position of flexion, before any attempt at extraction be made.

The Prague Method.—This is the quickest and simplest method of delivering the head, if it be in the pelvis ; it is not the best method when the head is above the brim. Standing at the patient's right side, the fingers of the left hand are hooked over either clavicle, and the feet are seized in the right hand. The first step insures flexion ;—the shoulders of the child are drawn directly upwards by the left hand, and detained in this position throughout the extraction ; by this means the pressure of the symphysis upon the occiput causes flexion of the head. With the right hand the body of the child is then swept forwards and upwards over the mother's abdomen. The head is thus made to roll out from behind the perinæum.

Martin's Method.—This method is suitable for all cases, whether the head be above or below the brim. Standing at the patient's right side, the

---

* This method, usually known as the Veit-Smellie method, I find is described in full by Smellie in vol. iii (M'Clintock's edition), case 303.

right arm is placed so that the body of the child lies upon it straddle-wise. As much as is required of the hand is introduced into the vagina ; the mouth is felt for, and two fingers introduced as far back as possible. This last precaution is necessary in order to avoid fracture of the jaw. With the fingers in the mouth the head is pulled down into a position of flexion. The left hand is then placed on the fundus, and by means of pressure upon the occiput the child is delivered.

Smellie's Method.—This is also suitable for any case. It is the most powerful method for extraction of the head which we have at our disposal. The right hand is used exactly as in Martin's method, whilst the fingers of the left hand are placed over the shoulders as in the Prague method ; but they are used for a different purpose, *i. e.* for traction. Flexion is obtained by jaw traction ; whilst to deliver, traction is applied both on the shoulders and the jaw. In this method, if the head be above the brim, we must first pull backwards and downwards, *i. e.* in the axis of the brim ; then directly downwards ; and then forwards, at the same time carrying the body of the child as it lies on the arm upwards over the mother's abdomen.

Forceps to the after-coming head only deserve a passing mention. They will extract it without doubt, but they are too slow. They require time for application, they are not always at hand, and no more power can be obtained by them than by Martin's or Smellie's method. It is surely better to accustom ourselves to the simplest method. If forceps be used, they must be locked under the body of the child, and traction applied in the axis of the pelvis.

Abnormal Mechanism in Breech Cases.—In a small percentage of cases, the chin of the after-coming head may rotate under the symphysis instead of posteriorly. It is then rather more difficult to deliver. There are two ways of doing so; the first is the better method. To perform it, draw the woman rapidly to the edge of the bed, in order to be able to depress the body of the child, then carry the latter well backwards. By this means the chin is drawn down from behind the symphysis. If the face do not follow easily, introduce the fingers into the mouth and apply traction, so that the face rolls out from behind the symphysis, the forehead following, and lastly the occiput. The other method is just the reverse of this. The body of the child is carried well forwards, so that the occiput rolls out over the perinæum, the forehead following, and lastly the face. The former method is to be preferred, as it enables us to reach the mouth, and so apply traction if necessary.

Prognosis.—The prognosis for the mother is no worse than in a vertex presentation; the fœtal mortality is, however, considerably higher. It is given variously as one in four and one in eleven. The longer the membranes remain intact, the better the os will be dilated, and the quicker will be the passage of the head through the pelvis.

FACE PRESENTATION.

Face presentations occur when the head becomes fully extended. For the reasons explained under the mechanism of vertex presentations, this condition is very rare (*v.* page 61); it occurs about once in 250 labours.

Ætiology.—Face presentations may arise in three ways. They may be due to :—

(1) Anything that prevents flexion; as,—tumours about the neck of the child, enlarged thyroid, and hydrothorax.

(2) Anything that tends to arrest the occiput at the brim, whilst at the same time permitting descent of the forehead; as,—obliquity of the uterus, contracted pelvis, or small tumours about the brim. In lateral obliquity of the uterus the head of the child, instead of being driven directly downwards into the brim, is driven against the opposite side of it; the result being, that the descent of the occiput is arrested, and the descent of the chin favoured.

(3) Malformation of the child's head; as,—a congenital dolichocephalic head. It is easy to understand that if such a thing as a congenital dolichocephalic head exist it would favour face presentation (*v.* page 61). We must, however, bear in mind, that an acquired dolichocephalic head is the result of the moulding which the head undergoes in a face presentation. So, what we imagine to have been the cause of a face presentation, may in reality only be the result of it.

Positions.—Two positions are recognised :—*first,*

when the chin is on the right; *second*, when the chin is on the left.

Mechanism.—The dimensions of the head that come into play in a face presentation are ;—the cervico-bregmatic, 3¾ inches; and the bi-parietal, 3¾ inches. The actual measurement of the cervico-bregmatic is almost the same as that of the sub-occipito-bregmatic, but it cannot be reduced to the same extent by moulding. The mechanism of face presentations resembles, very closely, that of vertex presentations, if extension be substituted for flexion, and flexion for extension. The various steps are as follow :—

(1) and (2) *Descent and Extension.*—The face engages with its cervico-bregmatic diameter in the transverse diameter of the pelvis. As the head descends, it, at the same time, extends, until the occiput is almost in contact with the back of the child.

(3) *Internal Rotation.*—Descent continues until the face reaches the pelvic floor, and then, obedient to the rule of internal rotation (*v.* page 63), the chin, which is lowest, rotates in front. It is characteristic of face presentations, that internal rotation occurs at a much later stage than in vertex presentations ; so much so, that the face may have appeared at the vulva before rotation commences.

(4) *Flexion.*—The chin now lies under the symphysis, and the head rotating round it is born; first the nose, then the eyes, forehead and occiput appearing from behind the perinæum.

Diagnosis.—A face presentation can be diagnosed by vaginal examination or by abdominal palpation.

By *vaginal examination* at the commencement of

labour we can ascertain nothing, as the face is detained for some time above the brim. If the membranes be not ruptured, we can feel the curious conical protrusion of them into the vagina. The cause of this protrusion is described under breech presentations; it occurs in all abnormal presentations (v. page 89). Later on, as the head descends, the presenting part can be felt ; but there is considerable difficulty is ascertaining what it is. As a result of the long labour, a large caput succedaneum forms upon the face, and causes it to resemble a breech. The diagnostic points are the supra-orbital ridges, the malar bones, and the mouth (which has to be distinguished from the anus, v. page 91). If there be still room for doubt, endeavour to pass a finger upwards, between the presenting part and the pelvis. In the case of a breech we come upon the angle of the groin, in the case of a face upon the ear. In examining a face presentation by vaginal examination, particular care must be taken not to injure the eyes. It is said also, that the introduction of the finger into the mouth may cause attempts at inspiration on the part of the child, and so lead to asphyxia.

By *abdominal palpation* a face presentation can generally be recognised. The first point that strikes us is the ease with which the limbs can be felt. This is due to the position in which the foetus usually lies ; its chest and abdomen are in close contact with the anterior uterine wall, and its back is directed posteriorly. Then, on making the pelvic grip, we feel, on the same side at which the limbs are, and resting upon the pelvic brim, a small tumour "like an animal's hoof" (Budin),—viz. the chin. On the other side, the pelvis is filled by a large tumour separated by a deep groove from the back of the child,—viz. the occiput. Also the occiput lies at a higher level in the uterus than the chin (v. Fig. 3). On the side at which the chin lies, the hand can be made to sink deeply into the pelvis, while on the opposite side it is prevented from doing so by the occiput.

Auscultation.—The fœtal heart is best heard over the limbs, instead of over the back as is usual in vertex or breech presentations. This is due to the

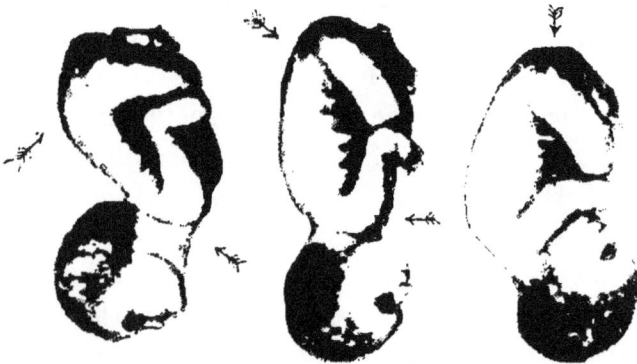

FIG. 10.—Schatz's method of converting a face into a vertex. The arrows show the directions in which the body of the child is pushed (Lusk).

fact, that, the chest and limbs of the child are pressed upwards against the abdominal wall, while the back is far away from it.

Treatment.—A face can be treated in two ways:—
(1) If recognised in time it may be changed to a vertex. (2) It may be allowed to remain a face and treated accordingly.

(1) If a face presentation be diagnosed in time

and it be decided to change it to a vertex, the
method of Schatz is the best means of doing so.
It requires, for its performance, three conditions to
be present :—

(a) Unruptured membranes.

(b) The face not yet fixed in the brim.

(c) A lax abdominal wall.

To obtain the last an anæsthetic is usually
necessary, though it need not be given if the
patient will refrain from straining. The details of
the operation are as follows :—Put the patient
under an anæsthetic ; palpate the abdomen carefully
in order to ascertain the position of the child ;
place both hands upon the shoulders and chest of
the child, and draw them directly upwards out of
the pelvis ; with one hand on the chest push it in
the direction of the child's back, while the other
hand on the breech pushes it in the opposite
direction ; lastly, push the breech directly downwards towards the pelvis, apply a tight binder, and
if the vertex do not fix, rupture the membranes (v.
Fig. 10). The danger of this method is, that flexion
of the head may not be complete, and a brow presentation result.

(2) If we elect to allow the face presentation to
remain, or if it be too late to alter it, it is well to
commence by warning the patient's friends, that it
will be a long and tedious labour, and that there is
considerable danger for the child. The treatment
of all abnormal presentations in the first stage, pre-
vails. Keep the woman in bed, and avoid anything likely to cause rupture of the membranes.
As the face approaches the perinæum, it is well to

examine, to see if the chin tend to rotate to the
front. If we think it is not doing so, forward
rotation can be assisted. Bear in mind the law
which governs internal rotation (*v.* page 63), and
press the forehead upwards during a pain. This
causes the chin to become the lowest part of the
face, and so favours its anterior rotation. This is
all that can be done to help the case. Forceps are
dangerous for the child and of slight use. They
will be discussed afterwards (*v.* page 174).

Abnormal mechanism.—In some cases, probably
owing to incomplete extension of the head, the
chin of the child rotates posteriorly into the hollow
of the sacrum. This is a practically hopeless condition for the child, unless either the head be very
small or the pelvis very large. The treatment is
mainly prophylactic. Endeavour to avoid its occurrence, as described above. If it occur, perforation is necessary.

Prognosis.—The mortality in face presentations is
considerably higher for both mother and child, than
in vertex presentations. Winckel gives 6 per cent.
as the maternal, 13 per cent. as the foetal, mortality.

CHAPTER X.

UNNATURAL PRESENTATIONS.

Brow Presentation: Proportion—Ætiology—Positions—Mechanism — Diagnosis — Treatment. Transverse Presentation: Ætiology—Mechanism ; Spontaneous Evolution, Spontaneous Version, Corpore Conduplicato—Diagnosis—Treatment.

UNNATURAL presentations are those presentations which cannot deliver themselves, without changing into another presentation. They include brow presentations, and transverse or oblique presentations.

BROW PRESENTATION.

The fœtus is said to present by the brow, when the part of the head between the supra-orbital ridges and the anterior fontanelle, is the lowest. The proportion of cases in which a brow occurs is difficult to ascertain ; as many cases of brow presentations change of themselves, during delivery, into face or vertex presentations. The proportion of cases in which a brow is either recognised and changed, or remains unchanged, is about 1 in 1500.

Ætiology.—The causes of a brow presentation are practically the same as those of a face (*v.* page 101).

Positions.—Two positions are recognised ;—*first,* when the occiput is on the left ; *second,* when the occiput is on the right.

Mechanism.—The diameters of the head, which engage in the brim in brow presentations, are the occipito-mental (5¼ inches), and the bi-parietal (3¾ inches). In many cases the head does not enter the brim at all. If it enter, the occipito-mental diameter of the head lies in the transverse diameter of the pelvis. Four possibilities are then present :—

(1) With a small head or a large pelvis the brow may be born unchanged.

(2) The brow may become converted into a vertex.

(3) The brow may become converted into a face.

(4) The head may become impacted in the pelvis.

If, in a favourable case, the brow is to be born unchanged, internal rotation must take place in such a manner, that, the brow rotates under the symphysis. The head is then born by extension.

Diagnosis. — By *abdominal palpation*, nothing peculiar is noticed about the body of the child. On making the pelvic grip the head will be found to lie well above the brim, with the chin and occiput on the same level (*v.* Fig. 3). By *vaginal examination*, no presenting part can be felt at the commencement of labour, owing to the high situation of the head. The membranes are felt bulging in a cone-shaped tumour, as is usual in abnormal presentation. Later in labour, if the head descend, the presentation is characteristic. On one side of the pelvis is felt the anterior fontanelle, and the smooth frontal bone with its median suture ; on the other side the supra-orbital ridges, the hollows of the eyes, and the malar bones. The caput succedaneum forms over the prominences of the frontal bone. By *auscultation*, the heart is heard slightly to one or other side of the median line, according to the position of the back of the child.

Treatment.—The first thing to understand clearly is, that a brow presentation is never to be left uncorrected, if it can be changed. If it cannot be changed into either a vertex or a face, then it is best left to itself. Nature will frequently correct a case which we cannot. Forceps should never be used except as the last chance before perforation, and not even then if the forehead be posterior. A brow presentation can be corrected in three ways :—

(1) By completing flexion, *i. e.* by turning it into a vertex.

(2) By completing extension, *i. e.* by turning it into a face.

(3) By version and bringing down a foot.

(1) If the brow be free above the brim, or at any rate not too deeply engaged, it may be converted into a vertex by the following method. The operator, with one hand in the vagina, pushes the head upwards out of the pelvis, directing his force especially against the forehead so as to favour flexion. An assistant then presses the child's shoulders in the direction of its back, as in Schatz's method (*v.* page 105). Flexion is finally completed, either by pushing the occiput downwards into the pelvis by pressure through the abdominal wall, or by pulling it down with the vaginal hand, which has been passed above it. The head is then kept in this position by means of a tight binder, and the membranes ruptured, if this have not been already done.

(2) If the brow be too far down in the pelvis to be converted into a vertex, we may endeavour to

change it into a face. This is best done by pressing upwards at each side of the large fontanelle during a pain, as this will tend to cause the descent of the chin. It will probably be unsuccessful, except in those cases in which nature would have brought about the same end.

(3) Podalic version should if possible be adopted, if, after a vertex presentation has been obtained, the head return to its original position.

If, at any time, the child be found to be dead, perforation should be performed, unless the head is about to be born immediately of its own accord.

Prognosis.—The maternal and fœtal mortalities are considerably higher in this presentation than in either normal or natural presentations. The fœtal mortality is given by Winckel as 29·4 per cent., the maternal mortality as about 10 per cent.

TRANSVERSE PRESENTATION.

A transverse presentation, a cross-birth, or an oblique presentation, are the different terms applied to the presentation of the fœtus, when it lies in the uterus in such a way that neither pole presents. Strictly speaking, a transverse presentation only occurs when the fœtus lies with its head at one side of the uterus, and its breech directly opposite. Similarly, an oblique presentation occurs when the fœtal head or breech lies in one hypochondrium, the other pole being in the opposite iliac fossa. The term cross-birth includes both these varieties.

Ætiology.—Any condition which causes a variation from the normal shape of the uterus or pelvis will

favour a cross-birth. The principal of these conditions are ;—contracted pelvis ; large lax uterus ;
hydramnios ; twins ; placenta prævia ; and tumours
of the uterus, as myomata. Abnormalities in the
shape or size of the child, will also favour it ; as,—
a very large fœtus ; a very small or premature
fœtus ; tumours on the body of the child ; and double
monsters. The relative frequency of transverse
presentations is about one in 136 deliveries.

Mechanism.—Transverse presentations, like brow
presentations, must never be allowed to remain
unchanged, except in the case of very small or
macerated infants. They can deliver themselves
sometimes under these conditions. Delivery may
then occur in one of three ways :—

(1) By spontaneous evolution.
(2) By spontaneous version.
(3) *Corpore conduplicato.*

(1) *Spontaneous evolution* occurs as follows :—The
shoulder of the child is driven down into the pelvis,
and the corresponding arm prolapses out of the
vagina. The shoulder then becomes fixed under
the symphysis, while the back, acutely flexed,
gradually appears from behind the perinæum. The
breech follows, and the lower limbs ; the last part to
be born is the head and remaining arm.

(2) *Spontaneous version* occurs, when the presenting
shoulder gets pushed away from the os by strong
pains, and the head or breech takes its place.
Delivery is then usually rapid.

(3) Expulsion *corpore conduplicato* is an extremely
rare occurrence, and is only possible with a very
premature fœtus, or one which is in an advanced

condition of maceration. The shoulder which presents is driven down into the pelvis, closely followed by the head and the rest of the trunk; the head and chest thus descending together.

Diagnosis.—A transverse presentation can readily be diagnosed by *abdominal palpation*. At the commencement of labour the pelvic brim is found to be empty. The head is then felt at one side of the abdomen, and the breech at the opposite.

By *vaginal examination* no presenting part can be felt at first. The membranes protrude unduly into the vagina during a pain. If the case have become a neglected shoulder presentation, then the shoulder can be felt and some of the ribs; the arm is usually prolapsed into the vagina. It is recognised as already mentioned (*v.* page 91). To decide whether it be the right or left arm, imagine yourself shaking hands with it. If your hand lie palm to palm with it, with the thumbs together, it is the right or left, according as the hand you are examining with is right or left. It must not be forgotten, that, although in a neglected cross-birth the shoulder is practically always driven down into the pelvis, still at the commencement of labour any part of the child's body may present. The middle of the back may lie lowest, or a foot and a hand may come down together.

Treatment.—A cross-birth may be treated in one of four ways:—

      (1) Postural treatment.
      (2) External cephalic version.
      (3) Internal or bi-polar podalic version.
      (4) Embryotomy.

(1) *Postural treatment* is often sufficient in cases of slight obliquity of the fœtus. To be of service, the membranes must be unruptured, and no limb must be prolapsed through the os. We must first understand the fact on which the method is based. As the fœtus lies in its normal attitude in the uterus, the breech and lower limbs, taken together, are heavier than the head. Accordingly, if the patient be placed on her side, and the membranes be unruptured, the breech will fall towards that side, and at the same time the head will rise towards the opposite side. To bring this into practice;—if the child's head be lying in the left iliac fossa, we place the patient on her left side. As a result the breech falls to that side, and the head rises to the opposite, so coming over the brim.

(2) *External cephalic version* requires the same conditions as the previous method. It is to be used when the obliquity of the fœtus is too great to be corrected by postural treatment. It will not always be successful, as the child tends to slip back to its original position. It should, however, always be given a trial, if the case be seen in time, as, if successful, it gives the child a better chance of life. It is useless to attempt it until the patient be in labour, as, otherwise, the head would not fix, and the malpresentation would recur. The child is turned by external version until the head comes over the brim ; the membranes are then ruptured and the head held there until it fixes, or a tight binder is applied to keep it in its place.

(3) *Internal podalic version*, followed by the drawing down of a leg, unless directly contra-

indicated, must be adopted in all cases in which external cephalic version has failed, or where the necessary conditions for performing it are not present. Any form of version is contra-indicated in cases of neglected shoulder presentation, when a considerable portion of the child has been expelled out of the uterus, or when Bandl's ring is more than 2½ inches above the symphysis (Winckel). Version may also be impossible to perform, owing to the force with which the child has been driven into the pelvis. When the leg of the child has been drawwn don into the vagina, it is well to leave the further expulsion to nature, unless there be an absolute indication for immediate delivery. (For the methods of performing version, *v*. page 177.)

(4) *Embryotomy* is indicated in a neglected shoulder presentation :—

> (*a*) When podalic version is contra-indicated owing to the condition of the uterus.
>
> (*b*) When podalic version is impossible.
>
> (*c*) When podalic version is difficult and the child is dead.

Decapitation performed by means of a Braun's blunt hook is the best mode of procedure (*v*. page 193).

Prognosis.—The fœtal mortality is very high. About 33 per cent. of children alive at the commencement of labour are born dead. The maternal mortality is about 5·5 per cent (Winckel).

CHAPTER XI.

MULTIPLE PREGNANCY.

Varieties —Relative Proportions—Twin Pregnancy—Ætiology—
Diagnosis—Presentations—Course of Labour—Treatment—
Prognosis—Complications : Locked Twins, Entangling of the
Cords, Fœtus Papyraceus.

MULTIPLE pregnancy is the term applied to the
presence of two or more children in the uterus.
Twins occur once in 88 births, triplets once in 7820,
quadruplets once in 366,913, quintuple births have
been recorded.

Twin pregnancy may occur in two ways. Either
one ovum contains two nuclei, both of which become
fertilised; or two separate ova may become fertilised.
In the former case, the children · are of the same
sex ; there is but one placenta and one chorion, but
there are two amnions. In the latter case, the
children may or may not be of the same sex ; there
are two placentæ, two chorions, and two amnions.
It must not be forgotten, that two placentæ may
grow in such a position that their edges coalesce,
and so there may appear to be but one placenta.
The nature of these cases is shown by the fact that
there are two chorions.

Diagnosis.—The only absolutely certain way to
diagnose twins is for two observers to count the

fœtal heart at the same moment, and to find that their results do not correspond. If monsters be excluded, twins can also be diagnosed by palpating two heads, two breeches, two backs, more than two large parts (viz. a head or breech), and more than four limbs.

Presentations.—Abnormal presentations are more common in multiple than in single pregnancies. The following table represents the proportion of the different presentations :—

Two head presentations occur in 49 per cent. of twin pregnancies.					
A head and a breech	..	31	,,	,,	,,
Two breeches	::	8	:,	,,	,,
A head and a transverse	,.	6	,.	,,	,,
A breech and a transverse	,,	4	,:	::	,.
Two transverse	,,	0·35	,.	,,	,,

(Spiegelberg.)

The usual course of labour is, that, after the birth of the first child, comes the second child ; then the placenta of the first child ; and then the placenta of the second child. In a small proportion of cases the first child is followed immediately by its placenta; then the second child and its placenta.

Treatment.—Having diagnosed the presentation of the first child, there is nothing further to be done but to allow it to be born naturally. Then palpate the presentation of the second child, as it may be lying transversely. If so, correct the presentation. Rupture the membranes of the second child about thirty minutes after the birth of the first, if they have not already burst spontaneously. This is always necessary, otherwise the second child might be retained in the uterus for some hours or even some days. Twins are frequently premature, and when the over-distended uterus has been relieved, by the birth of one child, it may lose its irritability. Cases have been recorded in which the second twin has been retained for a fortnight, or even more, after the birth of the first. In fact, some authorities recommend, in cases in which the placenta of the first child follows it, to put on a binder and keep the patient quiet, in hopes that the second child may not be born until full term. This treatment, however, exposes the woman to all the pain and expense of a second confinement. The object of waiting for thirty minutes after the birth of the first child, before rupturing the membranes, is to give the uterus time to regain its tone, and so minimise the danger of post-partum hæmorrhage.

Prognosis.—The prognosis in twin pregnancy is little worse, for the mother, than in a single pregnancy. For the children, the prognosis differs according to the presentation. But, as the children are usually small, and the maternal parts, at any rate for the second child, are well dilated, the mortality is less than the same abnormal presentation would cause in a single pregnancy. Many twins, though born alive, die during the first month of their existence, as a result of their premature birth.

Complications.—Dystochia may arise in twin pregnancy owing to the children becoming interlocked during birth. It may occur in three ways :—

(1) Two very small heads enter the pelvis at the same moment ; rotation is thus prevented, and further

delivery without assistance is impossible. The treatment is to endeavour to push up one head, so as to
allow the other to descend. If it fail to do so,
forceps must be applied. In very rare cases perforation may be necessary.

(2) Both children present by the head, one a little
in advance of the other. The head of the second
child becomes driven down into the neck of the
other, and so prevents any further descent. The
head of the second must be pushed up, and the
first extracted by forceps.

(3) The first child presents by the breech, and is
partially born. The second child presenting by the
head enters the pelvis in such a manner, that its
chin becomes locked under the chin of the after-
coming head of the first child. If the second head
cannot be pushed up, an attempt may be made to
extract it with forceps past the body of the first
child. If this fail, or if the first child be dead, it
should be decapitated, its head pushed up, and the
second child extracted by forceps. In any of these
cases in which embryotomy is necessary, the first
child should be sacrificed as it is the more likely to
be dead.

Entangling of the cords sometimes happens
in multiple pregnancy. As a result, one or both
children may die *in utero*. Also, during the birth
of the first child, the cord of the second may be
pulled down into a sharp angle, and so circulation
through it be prevented. Lest this accident happen,
the cord of the first child should never be pulled
upon.

A *foetus papyraceus* is formed when one child

dies *in utero*, but the other lives. As no air gains access to the dead child, it does not putrefy, but shrivels up and becomes mummified. The growing child then presses against it, and flattens it out against the membranes or placenta ; and it is found, after birth, adherent to them.

CHAPTER XII.

THE PUERPERIUM.

The Physiological Phenomena—The Involution of the Uterus—
The Lochia: Varieties, Amount—Lactation: Amount, Com-
position of Human Milk—The Condition of the Patient—
Treatment—Complications : Sub-involution.

THE puerperium is the term applied to the period
during which the woman is recovering from the
effects of pregnancy and childbirth. During the
puerperium the parturient canal is returning to its
normal condition and lactation is being established.

There are certain physiological phenomena to be
considered which are peculiar to this period. These
are :—

(1) The involution of the uterus.

(2) The lochial discharges.

(3) The establishment of lactation.

(1) The uterus takes six weeks to return to its
normal non-impregnated condition. Immediately
after delivery it weighs 24 ounces, and this has
to be reduced to the normal weight of 9 to 10
drachms (Heschl). This process of involution is
chiefly caused by the diminution that occurs in the
blood-supply of the uterus after delivery. The
uterus contracting tightly, compresses and so ob-
literates the greater number of its nutrient vessels.
The muscle fibres, as a result, undergo a fatty

degeneration, and come away in the lochia as fat drops. As the uterus involutes, it decreases in size ; and by the ninth day the fundus lies behind the symphysis (Winckel).

(2) The lochia is the discharge which comes away during the ten days succeeding delivery. It consists at first principally of blood. It also contains fragments of placenta, membranes, and decidua. Later on, cervical and vaginal epithelium, white corpuscles, crystals of cholesterin, and fat drops are found. From the third day onwards numerous forms of micro-organisms are present (Winckel). Three varieties of lochia are recognised. First comes the *lochia rubra* or *cruenta*, which lasts for about three days ; it is nearly pure blood. It gradually changes to the *lochia serosa*, which is sero-sanguineous. It persists until the sixth or seventh day, and then gradually passes into the *lochia alba* or *lactea*, which is principally mucus, and is creamy in colour, due to the presence of white corpuscles. The total amount of lochia that comes away is about 3¼ lbs., the lochia rubra composing far the largest portion of it.

(3) The fluid which is found in the breast during the first forty-eight hours, is called *colostrum*. It differs from ordinary milk, in that it contains less caseine, and more fat, sugar, and inorganic salts. The milk proper becomes established about the evening of the second day, and rapidly increases in amount. The quantity of milk which a woman secretes is shown by this table :—

Day		1	2	3	4	5	6	7	8	9	10	11
Amount in ounces		0	3⅓	7	8¼	13	15	17½	19	22	23	25

(*Winckel*).

The average composition of human milk at first
is :—

Albuminoids . . .	2·00
Fats	4·13
Sugar (lactose)	7·00
Inorganic salts . . . . ⸱ .	0·20
Water	86·67
	100·00

We can tell whether or not the patient be progressing favourably during the puerperium, by inquiring into certain points. They are :—

(1) The lochial discharges ;—their amount, their
colour, their odour, and the stain they leave upon the
napkin. Normal lochia should flow freely at first, and
cease gradually. Sudden stoppage sometimes corresponds to the onset of sepsis. The colour should
change according to the day of confinement, as
mentioned. If the lochia continue to be of a red
colour after the sixth day, it shows that some
degree of sub-involution is present. The normal
odour of lochia is heavy, and somewhat resembles
that of blood ; any putrid odour is pathological.
The stain on the napkin, caused by healthy
lochia, differs considerably from the stain caused by
putrid lochia. The former is red in the centre and
fades away towards the edge, which is colourless.
The latter is not so red in the centre, but becomes
deeper in colour towards the edge, which is clearly
defined.

(2) The normal temperature during the puerperium does not exceed 100·5° F. in the axilla. Any
rise above that, points to some abnormal condition.

(3) The pulse ranges between 50 and 90 beats

per minute. It is often a most important aid in the diagnosis of sepsis. If the temperature rise, but the pulse remain tolerably normal, the condition in all probability is not serious.

(4) The amount of sleep the patient has had, is a most important sign. Sleeplessness is often one of the first indications of commencing sepsis; and on the contrary, if the patient sleep well, she generally is progressing favourably.

(5) The aspect of the patient is also of the greatest importance. In any septic condition her face becomes drawn and pinched, and has a yellow tinge; the angles of the nostrils are drawn down; and the whole appearance is characteristic.

(6) The milk should flow freely, after the second day; a sudden cessation of it, points to septic infection.

The relative value of these points is well brought out in the following words. " If a patient with a high temperature look well, sleep well, and say she is well, she is at any rate not septic." " If a patient with a high temperature look very ill, sleep very badly, and say she feels very ill, she generally is very ill." " If a patient with a high temperature look very ill, sleep very badly, but say she is very well, she will probably die " (Smyly). The last condition is known as euphoria, and will be described under acute sepsis (v. page 272).

Treatment.—The treatment of the patient during the puerperium, is best considered under certain headings :—

Rest.—As soon as the patient has been delivered and comfortably settled, she should be kept perfectly quiet, and allowed to sleep if she can. During the first few days she will probably spend the greater part of her time in this manner. She should be kept in bed until at least the seventh day, and longer if possible. Whatever day she gets up, it must not be until the red lochia has ceased, as many cases of retroversion are caused by so doing.

Diet.—For the first two days the patient is kept on light nutritious food ; beef-tea, gruel, milk, tea and dry toast, and egg well beaten up,—anything of this nature may be given. After this the diet is more liberal, and, if her bowels have moved, she may have any ordinary digestible food. Stimulants are not necessary unless the patient be very weak. If she be in the habit of taking them regularly it may be advisable to continue them.

Bladder.—The patient should be made to pass water within six hours after her confinement. It is frequently impossible to get her to do so of her own accord. The recumbent position, and the presence of slight lacerations and bruises about the urethra, combine to prevent her. If this be so, the catheter must be passed for the first couple of days ; but, after this, every attempt must be made to induce the patient to pass water naturally. It is a very different matter to pass a catheter on the first or second day after confinement, and to pass it on the fourth or fifth. The lochia then contains bacteria, and may be lying decomposed about the external genitals ; a little of it carried into the bladder may start a severe cystitis. If the patient cannot pass water in the recumbent position, she should turn on her hands and knees, and endeavour to do so in this

position ; or, after the fourth day, she may get up and
stand by the edge of the bed. If, as a last resource,
the catheter must be passed, the patient must be
most thoroughly washed up. Then the urethra must
be exposed, and the catheter passed by vision, not
by touch. A silver or glass catheter should always
be used, and not a gum-elastic one, on account of the
difficulty of sterilising it.

Bowels.—The usually accepted idea is that the
patient should not get a purgative until the third
day. The majority of patients are, however, much
relieved by, and considerably the better for, a brisk
purgative on the evening of the second day. Castor
oil (ʒj—ʒij), Cascara sagrada (ʒj—ʒij), or Mag.
Sulph. (ʒss) may be given. A purgative should be
administered every second day during the puerperium, if the bowels do not move of themselves.

Complications.—The complications to be feared
during the puerperium are ;—hæmorrhage, sepsis,
and sub-involution. Of the two first I shall say nothing now.

Sub-involution is the resultant condition in those
cases in which the uterus does not diminish in size
in the normal manner. It is recognised by the per-
sistence of red lochia, and by the presence of the
fundus above the symphysis after the ninth day.
The immediate treatment is rest in bed, and the administration of ergot internally. Ergot given a few
times in tolerably large doses—up to a drachm of
the liquid extract—gives better results than small
doses long continued. If the condition persist,
the ultimate treatment consists in curetting, and the
correction of any malposition of the uterus which
may be present.

CHAPTER XIII.

ERRORS OF DEVELOPMENT, OR OF POSITION, OF THE PREGNANT UTERUS.

Malformations of the Pregnant Uterus—Malpositions of the Pregnant Uterus—Pathological Anteflexion—Retroversion—Incarceration of a Retroverted Uterus: Symptoms, Treatment—Prolapse of the Uterus: Treatment.

MALFORMATIONS OF THE PREGNANT UTERUS.

To understand these conditions, it is necessary to refer for a moment to the development of the genital

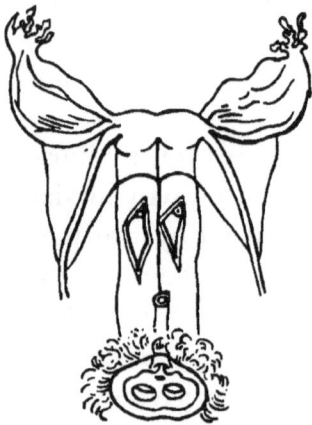

FIG. 11.—Double uterus and vagina.
(*Courty.*)

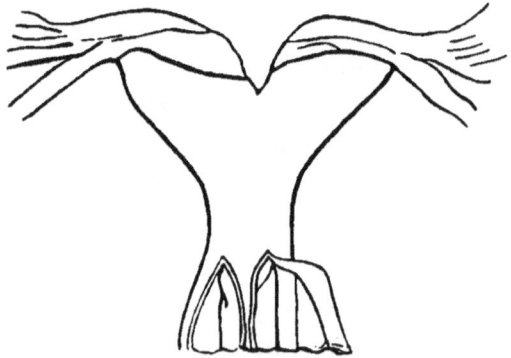

FIG. 12.—Uterus bicornis.
(*Schroeder.*)

organs. Two tubes—the ducts of Müller—run down at either side of the spine in the early embryo. They unite in their lower half, and the septum which at first separates them disappears. From the upper ununited portions of the ducts are derived the Fallopian tubes; from the lower portions which coalesce are derived the uterus and vagina. We see, then, that the Fallopian tube, with its corresponding portion of the uterus and vagina, was once a single duct. If this point be clearly grasped, it is easy to understand the different malformations that may arise :—

(1) The tubes may unite in their lower half, but may not coalesce, and thus a *uterus duplex* or *didelphys* be formed (*v.* Fig. 11).

(2) The tubes may not unite until the level of the cervix is reached, and thus a *uterus bicornis* may be formed (*v.* Fig. 12).

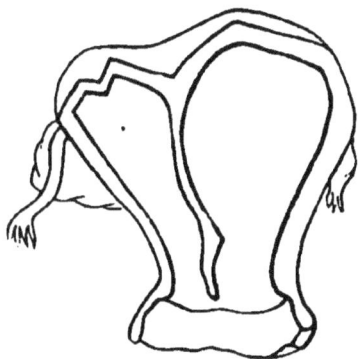

Fig. 13.—Uterus septus bilocularis.
(*Cruveilhier.*)

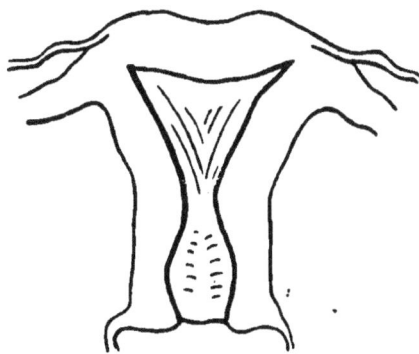

Fig. 14.—Uterus cordiformis.
(*Kussmane.*)

(3) The tubes may unite, but the septum may persist (*a*) in the uterus—*uterus septus bilocularis* (*v.* Fig. 13); (*b*) in the vagina—*vagina septa*.

(4) One Müllerian duct may develop whilst the

other remains rudimentary—*uterus unicornis* (*v.* Fig. 15).

(5) A depression may remain at the top of the fundus, corresponding to the point where the ducts united—*uterus cordiformis* (*v.* Fig. 14).

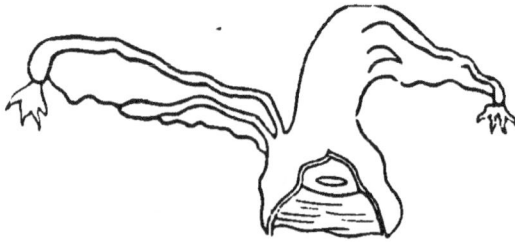

Fig. 15.—Uterus unicornis. (*Courty.*)

In order that pregnancy may occur in any of these abnormalities, it is necessary, that, at least, one side of the genital tube be fully developed.

If pregnancy occur in a double uterus, certain complications may arise :—

(1) Abortion. This is unusual.

(2) Tedious labour, due to the accompanying imperfect development of the uterine muscle, or to the obstruction offered by the other half of the uterus.

(3) Post-partum hæmorrhage, and retained placenta. This occurs particularly if the placenta be attached to the septum.

If pregnancy occur in the rudimentary horn of an uterus unicornis, the condition resembles extra-uterine pregnancy. The treatment is similar.

MALPOSITIONS OF THE PREGNANT UTERUS.

Pathological Anteflexion.—This condition occurs when the fundus is fixed in a position of anteflexion. It may be :—

(1) Congenital.

(2) The result of inflammation.

(3) The result of Makenrodt's operation of vaginal fixation of the uterus, for the cure of chronic retroversion.

If due to either of the first two causes, it usually gives little trouble. Frequent micturition may be the only symptom. In the last case the condition is more serious. Usually as the uterus increases in size it breaks free from its vaginal attachment, and then no harm results. If this do not occur, as the uterus grows, the cervix is gradually drawn upwards; the portion of the fundus which is attached to the vagina remains in front of, and below, the level of the cervix, while the posterior uterine wall develops sufficiently to accommodate the child. When labour comes on, the child's head is driven down into this cul-de-sac, instead of against the internal os. The cervix consequently does not dilate, and the uterus may rupture.

Treatment. — Dilate the os, if possible, with Barnes' bags, and apply forceps. If this cannot be done, it may be necessary to perform a Cæsarean section.

Anteversion.—This condition occurs in the later months of pregnancy in cases of contracted pelvis, and occasionally in pluriparous women. The uterus is pushed upwards, out of the pelvis, by the narrow

9

brim, and falls forward against the abdominal wall, a pendulous abdomen resulting.

Treatment.—Support the abdomen by an abdominal belt or binder.

Retroversion.—Pregnancy frequently occurs in a retroverted uterus, and then may terminate in four ways :—

(1) *Abortion.*—This happens very frequently. The retroverted uterus is subject to endometritis, which favours the occurrence of abortion.

(2) *Restitution.*—As the uterus increases in size, it gradually rights itself; and then pregnancy continues as usual.

(3) *Anterior development.*—This is a very rare occurrence. It is the reverse of the condition which may result from Makenrodt's operation. The fundus remains bound down in Douglas's pouch, but the anterior uterine wall develops sufficiently to permit the growth of the fœtus. There is thus at full term a cul-de-sac behind the cervix, into which the child's head is driven. The treatment in this condition is practically the same as in the case of an anterior cul-de-sac.

(4) *Incarceration.*—This is a serious condition, and, if not relieved (*i. e.* if the uterus be not replaced), will almost certainly result in the death of the woman. Its occurrence is favoured by the presence of a contracted pelvis, as the overhanging promontory prevents the uterus from rising.

Symptoms.—A tumour which is increasing in size fills the pelvis; the consequent symptoms are all the result of its presence. They are :—pain, constipation, and difficulty in micturition, all of which in-

crease from day to day. One day the patient
becomes unable to pass water, her bladder becomes
over-distended, and then the urine dribbles away in-
voluntarily (*ischuria paradoxa*). It is usually while
in this condition that she sends for medical aid.
On examination of the abdomen, a tumour is felt
extending up to the umbilicus, which yields a dull
note on percussion. This may put us off our guard.
We may think that the tumour is the uterus, while
in reality it is the over-distended bladder. On
making a vaginal examination, a tumour is felt filling
Douglas's pouch and pressing forwards towards the
pubes. The cervix is drawn upwards, and pressed
forwards, so that it lies above the symphysis, and
sometimes it may be impossible to feel it. The
urethra is also so drawn up, that it may be difficult
to find its orifice in order to pass a catheter.

Treatment.—The condition having been recognised, the first step is to empty the bladder. This
is sometimes a matter of great difficulty. If the
catheter cannot be passed in the ordinary manner,
place the patient in the knee-and-chest position, and
then endeavour to pass a flexible gum-elastic catheter. If this fail, the bladder must be tapped supra-
pubically. The next step is to replace the uterus,
under chloroform if necessary. If this can be done,
a pessary is inserted, and the condition is cured. If
repeated attempts fail, abortion must be brought on.
Owing to the position of the cervix, it may be impossible to pass any instrument into the uterus. If
all attempts fail, the uterus must be tapped with a
fine trocar through the posterior vaginal wall, and
a portion of the liquor amnii drawn off. This is a

certain method of procuring abortion, and is sufficiently safe if all due aseptic precautions be used.
If the condition be left unrelieved, death will be the
result. The uterine wall may slough from the continued pressure ; the bladder may rupture from
over-distension ; a very virulent form of cystitis may
result from retention of urine ; and consecutive
nephritis may result from the cystitis. Any of these
conditions, in turn, may give rise to septic peritonitis.

Prolapse of the Pregnant Uterus. — Pregnancy,
occurring in a uterus which is entirely prolapsed
outside the vulva, has been recorded. It is exceedingly rare. The usual condition met with, is one in
which the cervix protrudes out of the vagina. This
may be due to a hypertrophic elongation of the
cervix, accompanied by descent of the uterus, or
existing alone. The result of such a condition may
be serious. The exposed cervix becomes hypertrophied, its tissue dense and unyielding, and numerous
ulcers form upon it. When the patient comes into
labour the cervix may not dilate, and rupture of the
uterus may then result.

Treatment.—Replace the prolapsed uterus, and in-
sert a ring pessary. If the mucous membrane of
the cervix be dense and ulcerated, warm douches,
glycerine plugs, and hot baths will help to soften it.
If there be no inversion of the vagina, and the
cervix alone be prolapsed, Winckel recommends amputation of it in the early months of pregnancy. If
the case be seen too late for treatment, the patient
must be carefully watched when she comes into
labour. If the cervix do not dilate, it may have to
be incised, or even a Cæsarean section performed.

CHAPTER XIV.

DISEASES OF THE DECIDUA AND OVUM.

Composition of the Ovum—Hydrorrhœa Gravidarum—Hydatidiform Mole—Hydramnios—Oligo-hydramnios—Anomalies of
the Placenta and Funis.

THE decidua is the term applied to the hypertrophied and altered mucous membrane, which lines the
uterus during pregnancy, and in which the ovum
becomes implanted. The ovum consists from without inwards of;—the placenta; two membranes,—the
chorion and the amnion; the liquor amnii; the
umbilical cord; and the fœtus. Each of these structures may be the subject of pathological changes.

THE DECIDUA.

Hydrorrhœa Gravidarum.—This term is applied
to a collection of fluid which forms between the
membranes and the uterine wall. It is said to
be due to an inflammation, and glandular prolifera-
tion, of the decidua vera, involving also the decidua
reflexa (Lusk). The exact causation of the inflammation is little known, but it is probably one
of the many results of an antecedent endometritis.
The result of the inflammation is, that a watery

fluid becomes stored up behind the membranes. It remains there until it has filled all the available space, when it comes away with a gush. Several such gushes may occur during pregnancy. In some cases there is no further disturbance, but in others premature labour or miscarriage may result. The first attack may occur at any time after the third month. There is no treatment applicable to the condition during pregnancy. The patient should remain at rest in bed after the flow, especially if it be attended with uterine contractions. She should be warned that abortion or premature labour may occur. After pregnancy is over, and the uterus has returned to its normal condition, any endometritis present should be cured.

The Membranes.

Hydatidiform Mole.—This is the term applied to a peculiar myxomatous degeneration of the chorionic villi. It is also known as vesicular mole or myxoma chorii. The chorionic villi, instead of atrophying, proliferate, and become altered into little cysts which contain mucin. The fœtus dies, and is absorbed; while the cysts go on increasing in number, and finally fill the entire uterus. They vary in size, from that of a grape, to that of a pin's head. They are described as resembling a mass of white currants floating in red currant juice. The change commences before the end of the third month. If it happen to be a twin pregnancy, one ovum alone may be affected.

Varieties.—Two forms are described :—a simple form, in which the degeneration is confined to the ovum; and a malignant form, in which the cysts invade the uterine wall, and perhaps penetrate through it into the peritoneum.

Ætiology.—Little is known as to the direct causation of hydatidiform mole. It occurs particularly in multiparæ, and after chronic catarrh of the mucous membrane (Winckel); I have, however, seen a case in a primipara, and another in a 2-para. It is said to tend to recur in the subsequent pregnancies. As it starts in the chorionic villi it can only be the result of conception.

Symptoms.—The subjective and objective symptoms of pregnancy are present, but no fœtus can be felt nor fœtal heart heard, unless it happen to be a twin pregnancy. The uterus never corresponds in size to the period of pregnancy; it may be smaller, but is usually considerably larger. It feels more tense and more elastic than normal. There is a constant, blood-stained, watery discharge, in which small cysts may be found. Their presence is, of course, pathognomonic. Constant crampy pains also occur, due to the efforts of the uterus to expel the mass.

Terminations.—It is a very serious condition. If untreated one of four terminations may follow:—

(1) Spontaneous expulsion.

(2) Death of patient, from constant loss of blood.

(3) Death, from rupture of the uterus.

(4) Death, from peritonitis, caused by perforation of the uterus by the cysts.

Treatment.—Empty the uterus as soon as the condition is recognised. To do this, induce labour by dilating the cervix with Barnes' bags; the mass may then be expelled spontaneously. If this do not occur, introduce the finger, or the hand, and clear out the uterus thoroughly. There will be very free hæmorrhage, whilst this is being done; but, as soon as the uterus is empty, the bleeding will cease. Then douche out the uterus with hot creolin solution. Never curette in the first instance ; as, if it happen to be a malignant form of mole, the curette might perforate the uterus with great ease. Frequently, these patients will return after a fortnight or three weeks, on account of a recurrence of hæmorrhage. The uterus should then be curetted thoroughly, as bits of the mole are in all probability left behind.

The Liquor Amnii.

Hydramnios.—This is the term applied to an excessive amount of liquor amnii. The normal amount is about two pints ; anything over five pints is considered excessive, and the amount may even exceed twenty pints.

Varieties.—Two forms are described :—(*a*) Acute, coming on in a single night ; it is very rare. (*b*) Chronic, when the fluid accumulates gradually.

Ætiology—The pathology of hydramnios is very uncertain. It is found associated with syphilis of the child, anencephalus, spina bifida, and twins. It is said, by some, to be due to amniotitis ; but it is doubtful if such a condition ever exist.

Terminations.—(1) Premature labour may set in, as a result of the over-distension of the uterus.

(2) The uterus may rupture from over-distension.

(3) The patient may die of failure of the heart, due to the pressure of the enormous uterus.

(4) In less degrees of distension the patient may go to full term.

During labour many complications may occur. The first stage is tedious, due to the over-distension of the uterus, and consequent weakening of the muscle fibres. Malpresentations of the child are common. At the time of the rupture of the membranes, owing to the great rush of water, the child may be swept into a malposition, if it be not already in one, and the cord may become prolapsed. As a result of the sudden diminution in size of the uterus, the placenta may be detached, and hæmorrhage result. The second stage is precipitate, provided the presentation of the child be correct. The third stage, again, is tedious, owing to atony of the uterus; the placenta may be retained, and post-partum hæmorrhage result.

Symptoms.—The symptoms are those of pressure, on the abdominal and thoracic viscera, due to the over-distension of the uterus. Thus, we find:—constipation, from pressure on the rectum; frequent micturition, from pressure on the bladder; pendulous abdomen, from pressure on the abdominal walls; vomiting, from pressure on the stomach; dyspnœa and cardiac palpitation, from pressure respectively on the lungs and heart.

Diagnosis.—The uterus is considerably larger than it ought to be, in proportion to the period of pregnancy. The fœtus is felt with difficulty, and it may be impossible to hear the fœtal heart,—both, owing to the quantity of fluid which lies between the child and the uterine wall. By vaginal examination, nothing can be felt but the bulging membranes.

Treatment.—During pregnancy, support the uterus by an abdominal binder. Rarely, it may be necessary to induce premature labour, owing to the cardiac symptoms. When the patient comes into labour, do not allow the membranes to rupture spontaneously ; as most of the troubles that occur are due to the sudden rushing away of the waters. As soon as the os is as far dilated as is considered safe, introduce the hand into the vagina ; pass a couple of fingers upwards between the membranes and the uterine wall ; then slip a knitting needle or the stilette of a catheter along them, and puncture the membranes as high up as possible. The liquor amnii will then drain away slowly. Lastly, palpate the fœtus, to see if it be lying in a correct position.

Oligo-hydramnios.—In this condition the liquor amnii is insufficient in quantity. As a result, the amnion may become adherent to any part of the fœtus. As the latter grows, these adhesions are drawn out into bands, and these bands occasionally encircle the fœtal limbs, thus causing intra-uterine amputation, and similar accidents.

THE PLACENTA AND FUNIS.

Anomalies of the placenta in size, shape, or position, and of the cord in length and manner of insertion, are frequently met with. The most important of these are :—

(1) *Placenta membranacea.* — The placenta is large and greatly thinned out, so that it covers almost the entire internal uterine surface. Retained

adherent placenta may be thus caused, as the thin placenta crumples up when the uterus contracts, instead of becoming detached.

(2) *Placenta Succenturiata.*—The placenta, instead of being composed of a single mass, has several detached portions, only connected with the placenta proper by means of blood-vessels. These secondary placentæ are very likely to remain behind after delivery, and to cause post-partum hæmorrhage, or sapræmia. If they cause hæmorrhage immediately after delivery, they are discovered, if the uterus be explored. Sometimes they do not begin to cause symptoms until a day or two after delivery, when secondary post-partum hæmorrhage sets in.

(3) *Battledore Placenta.*—The cord is inserted into the edge of the placenta, instead of into the centre. It is of no clinical importance.

(4) *Placenta Prævia.*—The normal situation of the placenta is on the anterior or posterior uterine wall, with its lower border 2 to 4 inches above the internal os. If any part of it lie "so near the internal os, that it be torn off in the formation of the lower uterine segment," the condition is known as placenta prævia (Winckel). It will be discussed later.

(5) *Insertio Velamentosa.*—In this condition, the placental vessels which form the cord, do not unite upon the surface of the placenta, but run separately for some distance along the membranes. They are thus liable to be torn when the membranes rupture, and so to cause the death of the child.

CHAPTER XV.

ABORTION.

Ætiology—Varieties: Threatened Abortion, Cervical Abortion, Missed Abortion, Complete Abortion, Incomplete Abortion— Diagnosis of Abortion—Miscarriage.

ABORTION is the term applied to the expulsion of the ovum from the uterus before the formation of the placenta, *i. e.* before the end of the third month.

Ætiology.—The causes of abortion are divided into predisposing and exciting. The former are the more important, as they can be treated and cured in many cases. The latter are of slight importance, as they only tend to cause abortion if the predisposing causes be already present.

The direct predisposing cause in almost every case of abortion is endometritis, which may exist as a primary condition, or may be secondary to some other condition, as :—

(1) Malpositions of the uterus, especially retroversion.

(2) Bright's disease.

(3) Syphilis.

The exciting cause is usually some sudden or violent movement of the body, as, a fall, or a fit of coughing. Anything of this nature will readily bring on abortion, when the endometrium is unhealthy, but will never cause it in a healthy uterus.

A few exciting causes exist which apparently do not require any antecedent predisposing condition. Pyrexia, if it be sudden, and hyperpyrexia almost invariably, will cause abortion. Those fevers in which the temperature rises suddenly, as scarlet fever or smallpox, are more likely to have this effect than typhoid fever, in which the rise of temperature is gradual. The sudden elevation of temperature kills the fœtus, and the ovum is then expelled. Congenital syphilis of the infant acts in the same manner, *i. e.* it causes the death of the fœtus, which is then expelled.

Varieties. — The most satisfactory method of classifying the varieties of abortion is according to the treatment which they require. The following is for this reason a serviceable classification :—

(1) Threatened abortion :—(*a*) Those that do not require active treatment. (*b*) Those that require active treatment.

(2) Cervical abortion.

(3) Missed abortion.

(4) Complete abortion.

(5) Incomplete abortion.

Threatened Abortion. — When a woman, who is less than four months pregnant, commences to bleed, the hæmorrhage may be due to an extra-uterine pregnancy, or to a threatened abortion. The diagnosis between the two conditions will be discussed later. If it be a case of threatened abortion, there is probably more or less pain of a colicky nature, and, if a vaginal examination be made, the os may be found partially dilated.

Treatment.—According as the hæmorrhage is slight or is severe, so the patient will not or will require active treatment. The question of active treatment is entirely decided by the rate and strength of the pulse, and by the appearance of the patient. It is never possible to say whether an abortion be inevitable or not, unless a portion of it have left the uterine cavity. It is always possible to say whether a patient have lost as much blood as we consider safe. If it be a case which does not require active treatment, we endeavour to stave off the threatened abortion. With this object in view the patient should be kept at rest in bed, until all hæmorrhage and pain have ceased. Opium may be given to relieve the pain, and liquid extract of Hydrastis Canadensis to check the hæmorrhage. Ergot is not to be recommended. If the dose administered be large enough to have any effect on the hæmorrhage, it will also cause sufficient uterine contraction to increase the risk of the expulsion of the ovum. Hydrastis is said to act by causing contraction of the walls of the blood-vessels. At any rate, it does no harm, and is useful as a "placebo," if one must be given.

If, on account of the hæmorrhage, we believe the case to require active treatment, one of two methods must be adopted :—the ovum must be removed by the finger or a curette ; or the vagina must be plugged. These methods are not alternatives ; if it be possible to adopt the first we should do so ; if we cannot adopt it, the second method must be used.

If this rule be followed, we shall plug the vagina in somewhat less than one per cent. of cases of abortion requiring active treatment. It is possible to empty the uterus immediately, if the os will admit one finger, or even a curette. The former is to be preferred, as it removes the ovum more completely. Pass as much of the hand as is necessary into the vagina, and one finger into the uterus. Detach the

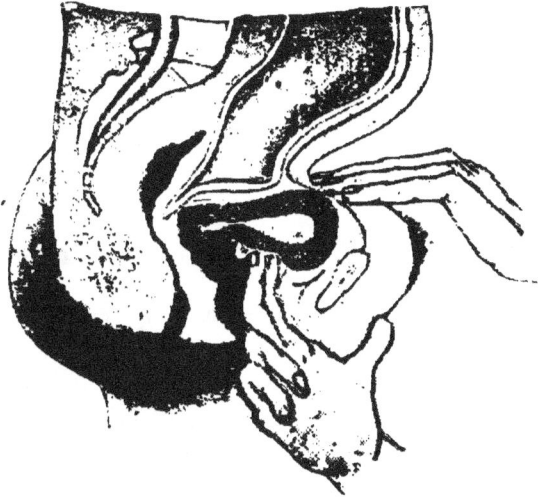

Fig. 16.—Bimanual method of expressing a detached ovum. (Semi-diagrammatic.)

ovum with the finger. Then take the latter out of the uterus, and place it under the fundus; i. e. in the anterior fornix, if the uterus be normal in position; in the posterior fornix, if the uterus be retroverted. Sink the other hand in the abdomen, and compress the fundus between the two hands (v. Fig. 16). The ovum is thus driven out of the uterus into the vagina. The uterus should then be well douched with hot creolin solution. If proper aseptic precautions have been used, the case will give no further trouble. If the os be not large enough to admit a finger, fix the cervix with an American bullet-forceps, and curette the ovum out with a Rheinstadter's flushing curette (*v.* Fig. 17). In the small proportion of cases in which the os is not large enough to admit even a curette, and the hæmorrhage is so severe as to require treatment, the vagina must be plugged, with the most careful aseptic precautions. The plug is left in for twelve to twenty-four hours, and then taken out. The os will then be found to be sufficiently dilated to permit the removal of the ovum with the finger, as

FIG. 17.—Rheinstadter's flushing curette.

described above. The dangers of plugging the vagina are considerable. Even if the plug itself be aseptic, still blood stagnates above it and putrefies. The decomposition then extends to the uterus; and, though the patient seldom actually dies as a result of this, she is frequently left an invalid for years, from tubal disease and pelvic peritonitis.

Cervical Abortion —This condition occurs when the ovum is detached from its situation in the uterus, and is expelled into the cervix. The external os does not dilate to allow it to pass, and the internal os contracts to some extent above it. It is thus retained in the cervix.

Treatment.—Incise the os externum bilaterally, and so make it sufficiently large to allow the passage of the ovum; then, express the ovum in the ordinary manner; and, lastly, stitch up both incisions. One stitch at either side is usually sufficient.

Missed Abortion. — An abortion is said to be missed when the ovum dies, but is not expelled. It may be retained in the uterus for some weeks, or even longer.

Symptoms.—The patient has obviously been pregnant. The uterus has increased to a certain size, but now has ceased to enlarge. The signs of pregnancy disappear, the uterus diminishes in size, and the breasts become flaccid. If the membranes rupture, the fœtus becomes putrid and causes a sanious discharge.

Treatment.—Dilate the cervix and empty the uterus with the finger or with a curette.

Complete Abortion.—A complete abortion consists in the coming away of the entire ovum. It requires no special treatment.

Incomplete Abortion.—An incomplete abortion consists in the coming away of any part of the ovum, the remainder being detained in the uterus.

Treatment.—As soon as the condition is recognised, turn the incomplete into a complete abortion, *i. e.* remove what is left behind. If the case be seen immediately after the portion of ovum has come away, and the os be still dilated, attempt to express the remainder of the ovum as directed above. If this fail, and if the os will admit the finger, introduce it; then detach the ovum, and express it. If this fail, or if it cannot be performed owing to the contraction of the os, curette the uterus carefully with

10

a blunt Rheinstadter's curette, having previously dilated the cervix if necessary. Never use a sharp curette in these cases, as it is very easy to curette away the soft muscle fibres of the uterus. Never plug the vagina in the case of an incomplete abortion, as decomposition is certain to occur above the plug. The expectant treatment of incomplete abortion is only mentioned to be condemned. It consists in waiting until one of three things happens :—

(1) The remainder of the ovum comes away. This is the most favourable termination, but it is not the commonest.

(2) The ovum decomposes.

(3) The patient loses so much blood, that it is considered inadvisable to allow her to lose any more.

If either the second or third termination occur, then and only then, is the uterus emptied. This is extremely bad treatment. It is much more dangerous to remove an ovum which is decomposed, than one which is not. Again, a woman, who is weakened by repeated hæmorrhage, is more liable to become septic, than one who has the normal quantity of blood in her body.

Diagnosis.—The diagnosis between ectopic gestation and abortion will be discussed in the next chapter. I am now considering only the diagnosis of the different varieties of abortion. To enable us to form this diagnosis two points must be attended to :—

(1) The nurse must keep everything that comes away from the patient *per vaginam.*

(2) The doctor must inspect such dejecta carefully,

with a view to discovering ;—(*a*) whether the case be one of abortion ; and, if so, (*b*) whether it be complete, (*c*) or incomplete. If nothing but blood come away, the case may be a threatened abortion, or it may be an extra-uterine pregnancy. If either a fœtus or chorionic villi be found among the discharged matter, it must be a case of abortion. If the whole ovum have come away, it is a complete abortion ; if only a part of it have come, it is an incomplete abortion. In many cases of abortion, unfortunately, everything that has come away has been thrown out by the patient's friends, or by the nurse. Then we have to rely on the history of the patient, and on the results of a vaginal examination. The former is unreliable, and, consequently, we must depend almost entirely upon the latter. Two points will then aid us :—

(1) The shape of the vaginal portion of the cervix.

(2) The continuance of hæmorrhage.

(1) The shape of the cervix varies, according to

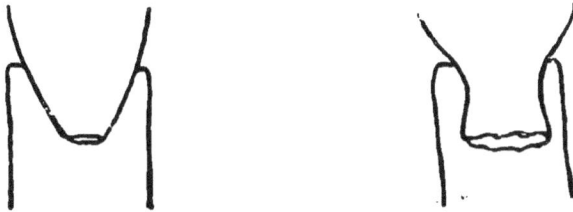

Fig. 18.—Diagram representing the shape of the cervix during, and subsequent to, the expulsion of the ovum.

whether the ovum be in the act of descending, or whether it have already been expelled. In the first case the cervix is cone-shaped, with the base of the cone above, *i. e.* in the region of the os internum. This is due to the presence of the ovum in the cervix ;—the os internum is dilated, the os externum closed (*v.* Fig. 18).   In the second case the cervix is cone-shaped with the base of the cone below, *i. e.* in the region of the os externum, and the apex above.   This is due to the fact that the os internum has closed again, whilst the os externum is still patulous.

(2) If the hæmorrhage have ceased and the os internum be contracted, the ovum has most likely been expelled.   If, on the contrary, the hæmorrhage continue, and, particularly, if there be a sanious discharge, some portion of the ovum must have been left behind.   In cases of doubt our treatment is governed by the symptoms.   If there be constant bleeding, the uterus must be explored, whether there be an ovum there or not.   If there be no hæmorrhage, and we do not know the exact condition present, it is better to curette.

Miscarriage.—A miscarriage is the term applied to the expulsion of the ovum after the placenta has formed, but before the fœtus is viable, *i. e.* before the twenty-eighth week.   These cases resemble full-term labour, and usually follow its course.   Before the fourth month, the ovum is universally attached to the uterus by vascular adhesions; accordingly, the detachment of any part of it causes free hæmorrhage.   The state of affairs differs in cases of miscarriage.   The placenta is formed, and is the only vascular attachment between the uterus and the ovum.   If uterine contractions occur, the os dilates, the membranes rupture, the fœtus is discharged,

and the secundines follow. Consequently, hæmorrhage is not a necessary accompaniment of these
cases.

Treatment.—The case is treated in the same
manner as a full-term labour. The fœtus is born,
we wait the usual time for the placenta to follow ;
if it remain behind, it is expressed or removed
manually, as may be necessary.

CHAPTER XVI.

EXTRA-UTERINE PREGNANCY.

Varieties — Primary Forms — Secondary Forms — Ætiology —
Before Rupture of the Tube : Symptoms, Diagnosis, Treatment — At the Time of Rupture : Symptoms, Diagnosis,
Treatment—After Rupture : Symptoms, Diagnosis, Treatment—Result of Fœtus remaining in the Abdomen—Table
showing the Varieties of Extra-uterine Pregnancy and their
Treatment.

EXTRA-UTERINE or ectopic pregnancy, is the term
applied to the development of the ovum outside
the uterus.

Varieties.—The following is the usual classification of the varieties met with :—

I. Primary forms	tubal	tubo-uterine or interstitial. isthmial. ampullar.
	ovarian (?)	
II. Secondary forms	from the interstitial	uterine. abdominal.
	from the isthmial	abdominal. broad ligamentous.
	from the ampullar	abdominal. broad ligamentous (very rare).
	from the ovarian	abdominal.

This   table   requires   slight   explanation.   The
primary   forms   are   classified   according   to   the
position   in   which   the   ovum   first   takes   root   and

grows (*v*. Fig. 19). The secondary forms are classified according to the locality into which the growing ovum extends, after rupture of the original site. Thus, a primary tubal pregnancy may, by the rupture of the tube, extend into the abdominal cavity or between the layers of the broad ligament, according to the part of the tube which ruptures, thus forming a secondary abdominal or secondary broad ligamentous pregnancy. Similarly, by the rupture of an ovarian pregnancy, if such a condi-

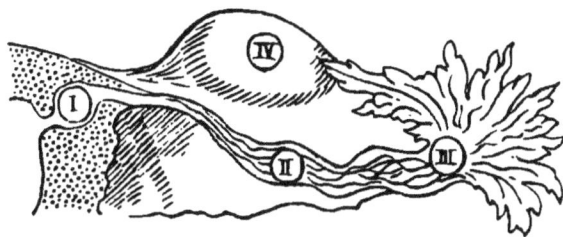

Fig. 19.—Diagram representing the various situations in which a primary extra-uterine pregnancy may develop : (1) interstitial; (2) isthmial; (3) ampullar; (4) ovarian. (Modified from ' The Norris Text-book of Obstetrics ').

tion occur, the ovum may extend into the abdomen, so giving rise, also, to a secondary abdominal pregnancy. In some cases, the ovum may be expelled by the contraction of the tube without its rupture, into the uterus in the case of an interstitial pregnancy, into the abdominal cavity in the case of an ampullar pregnancy. This termination is known as tubal abortion.

Ætiology.—The cause of ectopic pregnancy is a matter of much uncertainty. It is believed to be most commonly the result of gonorrhœal salpingitis, —a condition which frequently causes complete obstruction of the tubes, so leading to sterility. It is easy to understand how a lesser degree of inflammation may produce, not complete obstruction, but a partial obliteration of the lumen of the tube. This narrowing may be of such a nature, that it will allow spermatozoa to pass upwards, but will prevent the fertilised ovum from descending into the uterus. The same narrowing might exist in cases in which the obstruction had been complete; but, in which, through subsidence of the inflammation, a small lumen was again established. It is frequently to be noticed that an extra-uterine pregnancy follows on a lengthy period of sterility.

In discussing extra-uterine pregnancy, I shall refer to tubal pregnancy alone, as it is probably the only variety met with. I shall discuss the symptoms, diagnosis, and treatment under three heads :—

(1) Before,

(2) At the time of,

(3) After, rupture of the tube.

(1) **Before Rupture of the Tube.**   .

Symptoms.—The patient believes herself to be pregnant, and displays all the subjective and objective symptoms of early pregnancy. She has missed one or two monthly periods, and then slight irregular hæmorrhages occur. At the same time, she complains of cramp-like pains in the lower part of the abdomen. Most frequently a history of previous sterility can be obtained. On vaginal examination, a tumour is felt at one or other side of, or behind the uterus, apparently attached to one uterine cornu. It varies in size from that of a hen's egg, to that of an orange. It is unilateral, and is traversed by large blood-vessels, which can be felt pulsating through the vaginal fornix. The uterus is enlarged, and corresponds in size to the period of pregnancy.

Diagnosis.—The condition has to be diagnosed from a case of threatened abortion, complicated with a pyosalpinx or an ovarian tumour. The marked pulsation of the tumour, the fact that it is unilateral, and the history of the case, are the most important guides. A pyosalpinx is almost always bilateral.

Treatment.—Treat the case as if it were a malignant tumour, and remove it by abdominal section, or vaginal colpotomy, as may be thought best.

(2) **At the Time of Rupture of the Tube.**

Symptoms.—The first symptoms are those of violent internal hæmorrhage,—intense pain and sudden collapse. The pulse is feeble, usually rapid, but on the other hand it may be very slow. The temperature falls to 95° F. or 96° F. At the same time the uterus usually expels a false decidua, which has been formed synchronously with the growth of the ovum; and accompanying hæmorrhage occurs. The succeeding symptoms depend on what has happened, or is actually happening, inside the abdomen. The tube may rupture intra-peritoneally or extra-peritoneally. In the latter case the hæmorrhage usually is soon checked by the pressure of the tissues of the broad ligament, and the symptoms abate; very rarely profuse interstitial hæmorrhage may occur. The former case—intra-peritoneal rupture—is far more serious. Two terminations are then possible :—

(*a*) That diffuse hæmorrhage occurs into the abdominal cavity. This is rapidly fatal unless checked.

(*b*) That a retro-uterine hæmatocele is formed. This is the more favourable termination.

If a vaginal examination be made, at the time of collapse, nothing peculiar is felt. If the existence of a tubal tumour have been determined previously, we may be able to determine its disappearance. In cases in which the hæmorrhage is limited by adhesions, a retro-uterine tumour will be felt as soon as the blood has coagulated.

Diagnosis.—The diagnosis has to be made from a threatened, or incomplete abortion, for one of which it is almost always mistaken. The first point which should attract our attention, is the extreme disproportion between the condition of the patient, and the amount of *apparent* hæmorrhage which has occurred. The patient apparently has lost only a little blood, but she is anæmic, collapsed, with a feeble pulse, and a low temperature. Then, the decidua, which has been expelled, should be examined. No trace of chorionic villi or of a fœtus will be found.

If a hæmatocele form, it is most important to be able to recognise it. As felt from the rectum, it is a tumour which fills Douglas's pouch, boggy in consistency, and with a dome-shaped upper surface. It invests the rectum, and the uterus can be felt anteposed. It is the recognition of the fact that the uterus is anteposed, which distinguishes a hæmatocele from a retroverted pregnant uterus, for which it is most likely to be mistaken. If there be any doubt the sound should be passed, as the result of a false diagnosis would be disastrous. The treatment for a retroverted pregnant uterus is to replace it; whilst any attempt to move a hæmatocele would lead to fresh hæmorrhage, and perhaps directly cause the death of the patient.

Treatment.—This depends upon the nature of the case. If the patient be extremely collapsed, and diffuse hæmorrhage be occurring, she should be kept absolutely quiet, and stimulants administered. Usually a slight amount of clotting takes place over the ruptured vessels, the hæmorrhage ceases, and the patient rallies somewhat from the primary collapse. If no treatment be adopted, in all probability the clot will be dislodged, and a fresh hæmorrhage start, which will prove fatal. To avoid this, as soon as the patient has rallied in the least from the state of collapse into which she had fallen, open the abdomen, secure the bleeding vessel, and remove the ruptured tube. If the patient were operated upon, whilst in the condition of primary collapse, she would very probably die upon the table.*

If the case be not seen until a hæmatocele have formed, the correct subsequent treatment is still a matter of discussion. Some authorities recommend

* There is an increasing tendency amongst surgeons to operate on these cases the moment that they are diagnosed, and not to wait until the patient has recovered from the condition of primary collapse. By so doing, the risk that the hæmorrhage may not temporarily cease is avoided ; but, on the other hand, many cases are operated on under more disadvantageous circumstances than is necessary. The advisability of immediate operation in these cases, and of laparotomy in all cases of retro-uterine hæmatoceles, can, I think, be decided only by statistics.   .

that the abdomen be opened in all cases, and that
the hæmatocele be cleared out.   Others wait, on the
chance of the hæmatocele being absorbed aseptically,
and operate only if a rising temperature show that
suppuration is occurring.   This latter is probably
the better treatment, as by adopting it a considerable number of cases are saved the danger of any
operation.   On the other hand, if the hæmatocele
putrefy or suppurate, the dangers of an abdominal
section are very greatly increased.

(3) **After Rupture of the Tube.**

As I have shown above, certain consequences
may follow the rupture of the tube.   These are :—

(1) The patient may die, as a result of the
     hæmorrhage or of the operation.

(2) The ovum may die and be absorbed, or may
     be removed at the time of operation.

(3) The ovum may live and come to maturity.

The last case most frequently happens in extra-
peritoneal rupture of the tube.   It more rarely
occurs in cases of intra-peritoneal rupture.   It is
with this condition we are now concerned ; that
is, the symptoms and treatment of a case of extra-
uterine pregnancy, in which the ovum survives the
rupture of the tube.

Symptoms.—Until full term is reached, there
are no special symptoms to call attention to
the condition of affairs.   As soon as it occurs,
spurious labour sets in, the uterus expels a decidual
cast of itself, and the child dies.   The patient
notices nothing further for a few weeks, when she
may begin to think that she is past her proper
time for delivery.   She also notices that her

abdomen is smaller, a change which is due to the absorption of the liquor amnii. If the condition be not relieved, the abdomen continues to decrease in size, and the patient at the same time gradually becomes weaker. She suffers from various subjective sensations, such as a bad taste in the mouth, nausea, shiverings, and pains in the abdomen.

Diagnosis.—It is a very difficult matter to decide, for certain, in the later months of pregnancy, whether the ovum be contained in the uterus or in the abdomen. It is almost impossible to distinguish between the uterus and the extra-uterine ovum, owing to the distension of the abdomen; and there is an obvious objection to the use of the sound. It is said that the absence of the painless contractions of the uterus as felt by the hand, and of the uterine souffle, are points of importance. But inasmuch as the woman has no symptoms which call attention to her condition before the normal period of termination of pregnancy, she is seldom sufficiently carefully examined to bring into notice these small points. Consequently, the diagnosis is most usually not made until it is obvious that she is considerably past her normal time for delivery. Then, the diminished size of the ovum may allow the uterus to be felt as a separate tumour. The introduction of the sound determines the diagnosis; this is now permissible, as the child is dead, and the patient must be delivered, whether the pregnancy be intra- or extra-uterine.

Treatment.—The fœtus and placenta must be removed; the only question is, when shall the operation be performed? If the nature of the case be recognised before the death of the fœtus, are we to endeavour to save the child? The general opinion is, that it is better not to regard the life of the child in these cases, but to consider only the mother. Children developed outside the uterus are usually weak and likely to die, even if extracted alive; whilst the danger of the death of the mother from hæmorrhage, if the operation be undertaken at full term, is very great. The best rule to follow is this. If the case be recognised whilst the placenta is still small, i. e. in the fourth, fifth, or sixth month, operate at once. If, on the other hand, the condition be not discovered until near full term, it is better to wait for a month, or even two, after full term, and then operate. By this time the maternal blood-vessels, which supply the placenta, have diminished in size, and there is less risk of hæmorrhage. For the details of the operation itself, I refer the reader to one of the large text-books on obstetrics. It is one of the most difficult operations met with in abdominal surgery. The chief troubles are the difficulty of dealing with the placenta, and the separation of the numerous adhesions, which form, in some cases, between the child and the intestines. It suffices to say here, that, if possible, the placenta must be removed, as there is the gravest risk of its decomposing if left behind.

Cases of extra - uterine pregnancy which have advanced to full term, if untreated, may terminate in several ways. The child may undergo :—

(1) Mummification, calcification, saponifica-
     tion; or
(2) Suppuration.

If either of the terminations in the first group occur, the child may be carried by the mother in her abdomen for years. A *lithopædion* is the term applied to the condition that arises, when the membranes become the seat of calcareous deposits. The mother's health is always interfered with at first, probably owing to absorption from the dead child. Afterwards, as the fœtus becomes dried up, it only causes inconvenience by its size and weight. If the fœtus decompose or suppurate, the result is very different. A general suppurative peritonitis may start, and cause the death of the patient; or a localised abscess may be formed. In the latter case, the abscess bursts, either externally, or into one of the hollow viscera. It will continue discharging, perhaps for years, until either the patient dies of exhaustion or amyloid disease, or the entire ovum is discharged piecemeal. She may then recover, but such cases are rare.

The accompanying table may be of some use to the student, in understanding this complex subject (*v.* the next page).

An extra-uterine pregnancy is most frequently	If untreated, this condition will terminate, before the end of the third month, in	This termination may cause respectively	If this resultant condition be untreated it will terminate in	The best line of treatment consists in
Tubal.	(a) Intra-peritoneal rupture.	(a) Diffuse haemorrhage into the peritoneal cavity.	The death of the patient.	Immediate laparotomy, and ligature of bleeding vessel.
		(b) The formation of a retro-uterine haematocele.	(a) Aseptic absorption.	Rest in bed.
			(b) Suppuration.	Laparotomy and evacuation of the septic clots.
		(c) Slight haemorrhage, the ovum surviving.	Full-term extra-uterine pregnancy.	Laparotomy after term and removal of the foetus.
	(b) Extra-peritoneal rupture.	(a) Profuse haemorrhage into the tissues of the broad ligament extending sub-peritoneally.	The death of the patient.	Immediate laparotomy, and ligature of the bleeding vessel.
		(b) The formation of a haematoma.	Aseptic absorption.	Rest in bed.
		(c) Slight haemorrhage, the ovum surviving.	Full-term extra-uterine pregnancy.	Laparotomy after term, and removal of the foetus.

CHAPTER XVII.

INDUCTION OF ARTIFICIAL ABORTION AND OF PREMATURE LABOUR.

Artificial Abortion : Indications, Methods—Induction of Premature Labour : Indications, Methods—Version—Plugging the Vagina—Catheterisation of the Uterus—Intra-uterine Injection of Water or Glycerine—Dilatation of the Cervix—Rupture of the Membranes.

ARTIFICIAL ABORTION.

ARTIFICIAL abortion is the term applied to the induction of labour before the child is viable, *i. e.* before the twenty-eighth week. It is only justifiable under very exceptional circumstances.

Indications.—Abortion should be induced only in order to save the life of the mother. It is indicated in :—

(1) Cases of retroversion of the pregnant uterus, which cannot be replaced.

(2) In certain diseases of pregnancy, as hyperemesis; and, perhaps, in exceptional cases of cardiac, renal, or pulmonary affections.

(3) In cases of contracted pelvis, in which the child cannot be born alive. This last indication is not admitted by many good authorities, who hold that a Cæsarean section at full term, and not artificial abortion, should terminate the pregnancy. It is a moral, as well as a medical, question.

11

Methods.—Before the formation of the placenta (*i. e.* before the fourth month), dilate the cervix, and remove the ovum with the finger. The dilatation may be commenced by means of sea-tangle tents, and completed by means of Hegar's dilators. From the fourth to the sixth month, puncture the membranes with a stilette. From the sixth month induce labour by Krauze's method, as described under the "induction of premature labour" (*v.* page 164).

The Induction of Premature Labour.

Premature labour is the term applied to the onset of labour any time after the child is viable, but before full term. As the operation is usually performed in order to save the child's life, it is almost useless to attempt it before the thirtieth week. Theoretically, the child may be considered viable after the twenty-eighth week. Practically, the mortality among infants born before the thirtieth week is so high, that it is useless to induce labour, in order to save the child's life, before that time. Again, in cases of contracted pelvis, which is the most frequent indication, it is useless to induce labour after the thirty-sixth week. The child's head has reached its maximum size by this date, and labour induced after this will not procure an easier confinement, but will bring into the world a weaker child, than if the case be allowed to go to term.

Indications.—Premature labour may require to be induced for any of the following reasons :—

(1) Contracted pelvis measuring from 3 to 3½ inches in the *conjugata vera*.

(2) Habitual death of the fœtus at some period after it has become viable.

(3) Ante-partum hæmorrhage.

(4) Hydramnios causing urgent heart symptoms.

(5) Certain diseases of pregnancy, as hyperemesis; and, perhaps, in exceptional cases of cardiac and pulmonary diseases. Renal disease and eclampsia are added by many of the best authorities; in very rare cases the induction of labour may be advisable.

Methods.—There are many methods of inducing premature labour, both good and bad. No one method will suit every case. The method we adopt depends upon the indication for its adoption. I shall first give the various methods, and then discuss them. Premature labour may be induced, or is said to have been induced by :—

(1) Podalic version, followed by rupture of the membranes.

(2) Plugging the vagina.

(3) Catheterisation of the uterus.

(4) Intra-uterine injection of liquids, as glycerine or water.

(5) Dilatation of the cervix, digitally, or with Barnes' bags.

(6) Rupture of the membranes.

(7) Electricity.

(8) Vaginal douching.

(9) So-called ecbolics.

(10) Cupping the breasts.

(1) Version, followed by rupture of the membranes, is the method to be adopted in certain cases of

placenta prævia. It at the same time checks the hæmorrhage and induces labour, which are the two things we require. This method will be discussed in full later (*v.* page 177).

(2) Plugging the vagina is the method to be adopted in certain cases of accidental hæmorrhage. It acts in the same way as method (1) does in cases of placenta prævia, *i. e.* it induces labour and checks hæmorrhage. It will be discussed in full later (*v.* page 206).

(3) Catheterisation of the uterus. This is Krauze's method of inducing labour, and is perhaps the best method to adopt in cases of contracted pelvis. It is a very simple operation to perform, but is not free from risk, as it seems to be especially easy to infect the patient with sepsis whilst performing it. It is only under the most scrupulous aseptic precautions that it can be considered safe. To perform it, the patient is placed upon her back, in the cross-bed position, under an anæsthetic or not, as is thought best ; the parts are shaved and thoroughly washed, and the vagina well douched. The cervix is then exposed by passing a posterior speculum, and the anterior lip is seized and drawn down by a bullet forceps. Two, three, or even four flexible gum-elastic bougies are then passed, one by one, upwards between the membranes and the uterine wall, as far as they will go. They should be passed in very gently and allowed to take their own direction. If they meet with any resistance, withdraw them, and pass them again in another direction. The ends of the bougies which protrude are then wrapped round with iodoform gauze, in order to

protect the vagina. Labour may ensue in a few hours, or may not ensue for a few days. The bougies are taken out when the patient gets into strong labour, or when they have been in without result for twenty-four hours. In the latter case, after douching the vagina, a fresh set are passed in. If two or three sets have been passed without result, the os should be dilated by means of Barnes' bags ; labour will then ensue. The bougies used must be carefully sterilised. This is best done by boiling them for ten minutes, and then letting them lie for at least three hours in corrosive sublimate solution (1 in 500).

(4) Intra-uterine injection of water or other fluid —Cohen's method—brings on labour by separating the membranes from the uterine wall. Numerous deaths have been recorded, owing to the entrance of air into the veins, and also from shock.

Intra-uterine injection of a small quantity of glycerine—Pelzer's method—is more deserving of attention. Half an ounce of glycerine is injected slowly between the membranes and the uterine wall. It is said to be a very certain and rapid method, but also to be dangerous, as it tends to set up nephritis and to cause shock. Pelzer claims that it acts—

(a) By mechanical separation of the membranes.

(b) By direct stimulation of the muscle-fibres of the uterus.

(c) Owing to its hygroscopic properties, by drawing liquor amnii through the membranes and thus rendering them flaccid.

Quite recently Kossmann recommended and successfully practised the injection of 5 c.c. (85 minims) of glycerine into the cervical canal. He states that accidents are due to the using of large quantities for hygroscopic purposes, while small quantities used with the object of stimulating muscle-fibre, as in the rectum, are perfectly safe.

(5) Dilatation of the cervix is best practised in conjunction with Krause's method, if necessary. It can be performed digitally, or by means of Barnes' or Champetier de Ribe's hydrostatic dilators. Digital dilatation is very liable to tear the cervix, and is not to be recommended. Dilatation at all is very rarely necessary, and if it have to be performed it is best done by means of the dilators mentioned. Champetier's are probably the best form to use, as their action more nearly resembles that of the unruptured membranes.

(6) Rupture of the membranes—Scheele's method —is the most rapid method of inducing labour. As, however, the dilating action of the membranes is lost, labour thus induced is very tedious. It may be the best method to adopt in cases of hydramnios, as it will at once relieve the cardiac symptoms due to the pressure of the large uterus.

(7) Electricity is said to be of use, but is seldom attainable.

(8) Vaginal douching—Kiwisch's method—is almost useless, except in conjunction with other methods.

(9) Ecbolics are useless, unless administered in dangerous doses.

(10) Cupping the breasts is also useless.

CHAPTER XVIII.

FORCEPS.

Varieties of Forceps—Neville's Axis Traction Forceps—Methods of using the Forceps—Conditions necessary for the Application of Forceps—Indications for the Use of Forceps—Method of Application — Contra-indications to the Use of Forceps—Forceps in Occipito-posterior Positions of the Vertex, in Face Presentations, in Brow Presentations, in Breech Presentations.

The term " forceps," as used in midwifery, means an instrument adapted for seizing and extracting the head of the child.

There are two distinct varieties—(1) the short forceps ; (2) the long forceps. The short forceps is intended for extracting the head when it lies low down in the pelvis. It is not suitable in any other case. As the long forceps will deliver the head in any position, the short forceps has been given up, as being a needless addition to the obstetric armamentarium. The long forceps consists of two blades, an upper and a lower, each blade possessing two distinct curves,—a cephalic curve which enables it to be adapted to the child's head, and a pelvic curve which enables it to be adapted to the curve of the parturient canal. It is often a difficult matter for beginners to determine which is the upper, and which the lower, blade. To do so, imagine your patient in front of you lying on her left side. Then hold the blade in your hand, in such a position that its pelvic curve corresponds with the pelvic curve of the mother. If the concavity of the cephalic curve be then directed upwards, it must be the lower blade; if downwards, it must be the upper blade. There are many patterns of forceps, both plain and with an axis-traction apparatus. By an axis-traction forceps, we mean one that is so adjusted, that, by pulling in the direction shown by an indicator or by the position of the handles, we are always pulling the head

Fig. 20.—Neville's axis-traction forceps.

downwards in the true axis of the pelvis. Every practitioner prefers the pattern of forceps to which he is accustomed; but to any one who is buying his first forceps, I strongly recommend that known as Neville's (v. Fig. 20). It consists of Barnes' long forceps with Neville's axis-traction apparatus adjusted to it. The great advantage of this instrument is that the traction apparatus is entirely outside the vagina, when the forceps is applied; that it is absolutely uncomplicated; that it is a true axis-tractor; and that the forceps can be used with or without the traction apparatus as desired. In choosing one of these forceps the points that should be observed are :—

(1) That the arrow-head indicator be parallel to the fenestrated portion of the blades.

(2) That the forceps be not too flexible.

(3) That the portion of the blade which comes in contact with the child's head be flat, or even slightly concave, never convex as it is often made.

The forceps should be used as a tractor pure and simple. It is useless to use it as a compressor in order to diminish the diameter of the child's head. It seizes the head in the transverse diameter of the pelvis, approximately ; and, if it be made to compress the head, it causes a compensatory increase in that diameter of the latter which lies in the conjugate diameter of the pelvis, *i.e.* in the diameter which most requires reduction. The forceps also should not be used as a lever, *i. e.* by giving oscillating movements to the handles ; as, by so doing, the mother's soft parts are made the fulcrum, and considerable harm may be done.

Conditions necessary for the Application of Forceps. —Before the forceps can be applied in any case, four conditions must be fulfilled :—

(1) The greatest diameter of the head must have passed the brim.

(2) The membranes must be ruptured, and retracted over the presenting part.

(3) The os must be sufficiently dilated to allow the head to pass through, without any risk of causing laceration of the cervix.

(4) The bladder must be emptied.

Let us discuss these points.

(1) *The greatest diameter of the head,* &c.   The forceps is an unsuitable instrument for moulding the head.   It tends to elongate the diameters which require diminishing; and thus more force is required for delivery than would be necessary if the head were suitably moulded.   As a matter of fact, nature will mould a head through a pelvis through which it could not be extracted by the forceps. Even if it be extracted, the extra force which must be employed is a direct source of danger to both mother and child.

(2) *The membranes must be ruptured,* &c.   If the membranes be not ruptured, they are included between the forceps and the child's head; and, while we are extracting the head, we are also detaching the placenta.   This accident may result in severe post-partum hæmorrhage, and in the death of the child.

(3) *The os must be dilated,* &c.   It is practically never necessary to extract a head through an incompletely dilated os.   If it be done, laceration is almost certain to result, and we are never sure how far a laceration once started may extend.   Even if no harm be done in the immediate present, it leads, in the near future, to all the trouble and ill-health attending lacerated cervix and ectropion of the cervical canal.   If we must deliver by the forceps through a semi-dilated os, as may be necessary in rare cases of prolapse of the cord, the cervix should be incised bilaterally, and thus laceration avoided.

(4) *The bladder must be emptied.*   The distended bladder takes up a certain amount of room, and so compels more force to be used for extraction.

Indications.—The indications for the use of the forceps may be classified as :—

    (*a*) Indications on behalf of the child.

    (*b*) Indications on behalf of the mother.

In the first group are included :—

    (1) Fœtal heart-rate rising above 160, or falling below 120.

    (2) Tumultuous movements on the part of the fœtus.

    (3) The coming away of meconium, unmixed with liquor amnii, in a head presentation.

    (4) Prolapse of the cord (*v.* page 257).

In the second group are included—

    (1) Hæmorrhages of all kinds (*v.* pages 207, 211).

    (2) Threatened rupture of the uterus (*v.* page 232).

    (3) Prolonged second stage, *i.e.* lasting more than four hours.

    (4) Convulsions (*v.* page 265).

    (5) Cardiac, pulmonary, and renal disease.

    (6) Hæmatoma of the vulva (*v.* page 283).

These are indications for forceps, only if the conditions mentioned above be fulfilled.

Method of Application.—The forceps may be applied with the patient on her side, or on her back. The former position is to be preferred, as it is easier to observe the progress of the case, the perinæum can be better protected, and only one assistant is required. If the case require considerable force to be used, then the patient may be turned on her back, as more power can be obtained when she is in this position.

With the patient on her side, the forceps is applied as follows :—Place the patient fully under an anæsthetic, wash thoroughly the external genitals, pass a catheter, and douche out the vagina. Then take the lower blade in the right hand, using soap as a lubricant (*v.* page 3). Pass as much of the left hand as possible into the vagina. If the cervix can be felt, slip the fingers between it and the child's head. If it cannot be felt, it is retracted and obliterated, and, therefore, there is no fear of including it between the forceps and the child's head. Enter the blade at right angles to the symphysis, and then pass it upwards along the palm of the left hand, carrying the handle at the same time towards the mother's right thigh. Use little or no force, but allow the blade to take its own direction. If it catch in anything withdraw it slightly, and then pass it up again. Above all, remember to which side you want the blade to go ; and that when it is there, the handle will lie at the opposite side. When the lower blade is in position, pass the other in the same manner, with the exception that the handle is carried towards the left thigh. Then lock the blades. If there be any difficulty in doing so, do not force them to lock, but withdraw one and pass it again. If the blades absolutely refuse to lock, it is probably not a case for the forceps. The administration of chloroform may be ceased as soon as the blades are locked. Having adjusted the traction apparatus, apply traction in the direction shown by the indicator. Pull intermittently, and during the uterine contractions, if there be any. As soon as the head reaches the perinæum, allow it to dilate the

latter for a moment then to recede, and repeat this
manœuvre several times before extracting; as, by so
doing, the perinæum is more fully dilated. Deliver
the head as has been previously described under
normal labour (*v.* page 73), taking off the blades
before it is born, or not, as is preferred. Anæsthesia should be over by the time the head is
born. The rest of the extraction calls for no further
remarks.

If the forceps be applied with the patient in the
dorsal position, the only difference lies in the fact
that the lower blade is passed with the left hand,
the right hand being in the vagina.

If Tarnier's or Milne Murray's forceps be used,
we must apply traction, in such a manner, that the
rods of the traction apparatus are always in contact
with the handles of the forceps. If forceps without
any axis-traction adjustment be used, we must pull
so as to suit the curve of the parturient canal.
That is, if the head be at the brim, first, backwards;
then, downwards and backwards; then, downwards;
downwards and forwards; and, finally, almost straight
forwards.

Contra-indications to the Use of the Forceps.—The
forceps is contra-indicated in two classes of cases :—

(1) In secondary uterine inertia, on account of the
danger of post-partum hæmorrhage (*v.* page 226).

(2) In contracted pelvis if the head have not
passed the brim, because the forceps is not a suitable instrument for moulding the head (*v.* page 170).

I do not mean by this that the forceps is never
to be used in these cases; it may be necessary to
apply it. I mean that it is only to be used as

a last resource, and is not to be applied merely because the mother is more than four hours in the second stage.

Extraction by the forceps in *occipito-posterior positions* of the vertex is always more difficult, and requires the exertion of more force than is necessary in a normal position of the head. During extraction the head may rotate, so that the occiput comes to lie beneath the pubes. If this happen, the forceps must be removed and reapplied in its corrected position. If the occipito-posterior position persist, the forceps must be carried well forward over the mother's abdomen, as the head is being delivered, in order that the occiput may sweep over the perinæum. The latter is very frequently torn in these cases.

In *face presentations*, the forceps is of little use. If it be applied whilst the long diameter of the face lies in the transverse diameter of the pelvis, it will in all probability slip. If it do not slip, the pressure of the blades on the neck of the child will probably kill it. After rotation of the face has occurred, the head is so nearly born that the forceps is unnecessary (*v.* page 106).

In *brow presentations*, the forceps is contra-indicated, unless the brow change into a vertex. The brow is much more likely to become corrected, into a face or vertex, if left to itself, than if the forceps be applied (*v.* page 109).

In *breech presentations*, the forceps is liable to slip and also to harm the child. It is better to extract an impacted breech by other means (*v.* page 182).

CHAPTER XIX.

VERSION—IMPACTED BREECH—IMPACTED SHOULDERS.

Version : Varieties—Cephalic Version : Indications, Methods, External Version, Bipolar Version—Podalic Version : Indications, Methods, External Version, Bipolar Version, Internal Version—Contra-indications to Version—Impacted Breech : Diagnosis, Treatment—Impacted Shoulders : Treatment.

VERSION.

VERSION is the term applied to the operation by which the presentation of the child is changed. There are two varieties of version, named after the resultant presentations :—

(1) Cephalic version.

(2) Podalic version.

Cephalic Version.—The operation of cephalic version consists in changing the original presentation of the child into a head presentation.

Indications.—Cephalic version is indicated in faulty presentations of the child, if the following conditions be fulfilled :—

(1) If rapid delivery be not required.

(2) If there be nothing to prevent the child's head engaging in the pelvis.

(3) If the presenting part be not fixed.

Methods.—It can be performed by (*a*) external

manipulation—Wigand's method ; or (b) by combined external and internal manipulation—Braxton Hicks' method.

(1) To perform external version we require a lax abdominal wall, and unruptured membranes. If the patient strain, she must be placed under an anæsthetic. As soon as this is accomplished, ascertain by palpation the exact position of the child ; and, by a series of pushing movements, press the head in whatever direction will bring it over the pelvis by the shortest route, at the same time pressing the breech in the opposite direction. Then, if the os be nearly dilated, rupture the membranes ; and either hold the head over the brim until the uterine contractions cause it to fix, or pin a binder tightly round the patient's abdomen, which will have the same effect. It is useless to turn the child before labour has commenced, as it would slip back into its original position.

(2) Cephalic version by the combined method of Braxton Hicks may be performed soon after the membranes have ruptured. To perform it, introduce as much of the hand into the vagina as is necessary, and push the presenting part upwards out of the brim. Then, with the other hand on the abdominal wall, press the head down, and ensure its remaining there by the same means as in external version.

Podalic Version.—This is an operation which is far more frequently required than cephalic version. It consists in changing the original presentation of the child into some variety of pelvic presentation, and most frequently into a footling presentation, by drawing down a foot.

Indications.—Podalic version is indicated in a variety of conditions, viz. :—

(1) In certain cases of malpresentation of the head, *i. e.* face and brow presentations (*v.* page 110).

(2) In certain cases of prolapse of the cord (*v.* page 256).

(3) In most cases of placenta prævia (*v.* page 210).

(4) In certain cases of contracted pelvis, as recommended by some authorities (*v.* page 248).

(5) In cross-births, in which cephalic version has either failed, or cannot be performed (*v.* page 113).

Methods.—Podalic version can be performed by :—

(1) External manipulations only, if it be not necessary to bring down a foot.

(2) Combined internal and external manipulations.

(3) Internal manipulations.

(1) External podalic version can be performed under the same conditions as cephalic version, namely,—lax abdominal walls, the presenting part unfixed, and unruptured membranes. It is performed in exactly the same manner as cephalic version; except that the breech, instead of the head, is brought over the pelvic brim.

(2) Combined or bi-polar version, after the method of Braxton Hicks, is principally indicated in the case of placenta prævia. To perform it we require lax abdominal walls, unruptured membranes, the presenting part not fixed, and an os which is sufficiently dilated to admit at least two fingers. An anæsthetic is almost always necessary. Place the patient in the cross-bed position, ascertain by palpation the exact position of the child, and then turn the child

12

by external version into a transverse presentation.
The child must be turned in such a direction, that
its back will be towards the fundus uteri, and its abdomen facing the pelvic brim. If this be done, the
foot is lying in the neighbourhood of the internal os.

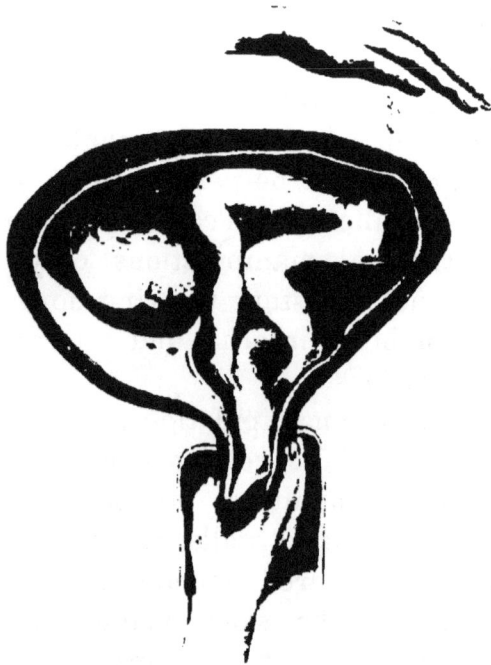

FIG. 21.—Method of completing bi-polar version in a case in which
the size of the os will not permit of the presence of the foot and
the two fingers at the same time. The hand in the vagina pushes
the cervix upwards while the foot is made to descend by pressure
upon the breech (diagrammatic).

Then introduce the whole hand into the vagina
and two fingers into the cervix, rupture the membranes, and with the hand on the abdomen press
the breech downwards ; the foot can then be seized

and brought out into the vagina. This is easily
accomplished if the os be fairly well dilated. In
some cases, however, the os may be quite large
enough to admit two fingers by themselves, or to
allow the foot to descend by itself, but it may not
be large enough to allow all three to pass through
it at the same time. If this be so, proceed as follows.
Having seized the foot with the fingers in the uterus,
draw it down until the toes are through the os internum. Then withdraw the fingers into the vagina,
and attempt to push the cervix upwards over the
foot. At the same time press upon the breech
through the abdominal wall, so as to cause the foot
to descend (*v*. Fig. 21). When half the foot has
by this means been brought into the vagina, seize
it and draw it downwards. Lastly, with the abdominal hand push the head up to the fundus.

(3) Internal podalic version can be performed :—
if the os be sufficiently dilated to admit the entire
hand into the uterus ; if the presenting part be not
too firmly fixed to be displaced ; and, if no contra-
indication to version exist. It most frequently
requires to be adopted in cases of neglected
shoulder presentation. It can best be performed
with the patient upon her back. Commence by
ascertaining the exact position of the child by
palpation, then empty the bladder, and wash the
patient. In a transverse presentation, introduce
that hand into the uterus which corresponds to the
side at which the limbs are,—*i. e.* if the limbs be on
the right side of the mother, use the right hand, and
vice versâ. In a vertex presentation, the right hand
is to be preferred for every position of the child,

except when its limbs are to the right and in front ; then the left hand is more suitable. If the operation have to be performed with the patient upon her side, place her upon the side at which the limbs are, whatever the position of the fœtus, and use the opposite hand.

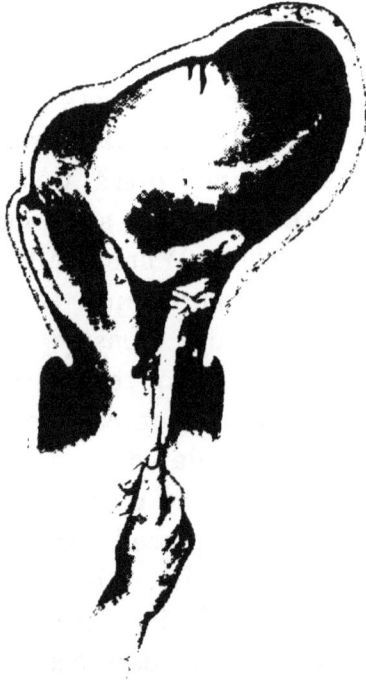

FIG. 22.—Method of completing a difficult case of internal version, by means of a gauze fillet (diagrammatic).

Having introduced the proper hand, seize the first foot that can be felt, and draw it down-wards into the vagina. The child is now lying with both head and breech in the lower part of the uterus. The last step of the operation consists in pushing the head up into the fundus, while at the same time we draw the foot down deeper into the vagina. This may be very easy, or it may be extremely difficult, or even impossible, according to the degree of force with which the uterus has contracted down upon the child. If the head cannot be made to rise to the fundus, as I have described, a simple and most successful expedient is as follows: —Tie a strip of iodoform gauze, sufficiently long to extend outside the vulva, to the foot which is in the vagina. Seize the strip with one hand, outside the vagina, and pull upon it; while, at the same time, the other hand in the vagina pushes the head upwards out of the pelvis (v. Fig. 22). If this fail, embryotomy in some form will be necessary, as the child cannot be delivered in its present position.

If an arm be prolapsed into the vagina, pay no attention to it at first, but draw down the foot as directed. The arm will be found to slip up, as the head ascends to the fundus.

Difficulties in the performance of internal version are often manufactured in two ways :—(1) by not having ascertained the exact position of the child at the commencement ; (2) by passing the hand outside the membranes instead of inside.

Contra-indications to Version.—Version is altogether contra-indicated by the presence of certain conditions :—

(1) If the previous contractions of the uterus have been so strong that the fœtus is in great part expelled from its cavity. In order to turn, the expelled portion of the fœtus would have to be replaced in the uterus, and there is not room for this.

(2) If it be obvious that the child cannot be

delivered, without embryotomy or craniotomy, even after version. The perforation of the after-coming head is a most difficult operation.

(3) If the membranes be long ruptured, and Bandl's ring be more than 2½ inches above the symphysis (Winckel). In this case rupture of the uterus would most probably result.

IMPACTED BREECH.

By the term impacted breech, we mean the condition that occurs when a breech presentation is driven down into the pelvis, but cannot advance further. It is usually due to an exceptionally large breech, or to the failure of a normally-sized breech to rotate.

Diagnosis.—An impacted breech is diagnosed, when the breech remains in the pelvis, and makes no progress, although the pains be strong.

Treatment.—First endeavour to express the breech by firm pressure upon the fundus, during the pains. If this fail, endeavour to bring down a leg. In some cases it is possible to do this, but in a considerable number it is impossible, as the breech is too low in the pelvis. If we decide to attempt to bring the leg down, place the patient in the cross-bed position. Introduce the entire hand into the vagina, and slip two fingers upwards along the posterior thigh. If the child be lying with its knees flexed, the foot will be found near the buttock, and can be seized and drawn down. If, on the other hand, the knees be extended, slip the fingers still further upwards along the thigh, until the knee be reached ; and, then, by pressure upon the anterior aspect of the lower leg just below the knee, the leg is made to flex upon the thigh, and so brought down. If the leg be got down, it diminishes the size of the presenting part, and gives a point upon which to apply traction. We must take special care to get below the knee, before endeavouring to bring down the leg, otherwise there is great danger of fracturing the femur.

If the leg cannot be brought down, we must resort to traction upon the breech. Place the patient in the cross-bed position, endeavour to get two fingers into the angle of the anterior groin, and then apply traction. By this means the breech is brought sufficiently low to enable us to pass the fingers into the posterior groin. So, by pulling alternately on one, and the other, the child is extracted. The strength of the fingers, which are used to make traction on the groin, can be greatly increased by grasping the wrist firmly, during the traction, with the other hand.

If the impacted breech still resist our efforts, endeavour to pass a fillet of iodoform gauze over it. There are several instruments for accomplishing this, but an ordinary gum-elastic catheter, with a stilette, will answer every purpose. Thread the catheter with a piece of boiled silk or twine, and then pass in the stilette. Bend the top of the catheter so as to form an arc of a circle, similar in size to the girth of the child's thigh. Then slip the catheter upwards, along the anterior vaginal wall, until the end of it can be turned inwards over the groin. Now, holding the instrument by the ring of the stilette, push the catheter itself upwards. The curve, which has been given to the stilette, will direct the catheter so that the point comes down between the thighs. The end of the twine with which it is threaded is then seized, and the catheter withdrawn. Lastly, a piece of iodoform gauze is tied to the twine, and, by means of it, drawn over the groin. This furnishes a soft but strong fillet, with which the breech can be drawn down. Traction should always be made in the axis of the pelvis, and care be taken that the fillet comes well down into the angle of the groin, and does not lie against the femur, as fracture would then be very likely to occur.

If the child be dead, and extraction difficult, a cephalotribe may be applied to the breech; or, if one be not to hand, a pair of forceps tightly screwed up may be used instead. A blunt hook is a most dangerous instrument for extraction. Even in skilled hands it may break the femur of the child, or tear the femoral vessels; whilst, in unskilful hands, much damage may be done to the mother's soft parts. As soon as the breech of the child has passed the vulva, the case is managed in the same manner as an ordinary breech presentation.

IMPACTED SHOULDERS.

Impacted shoulders occur sometimes after the birth of the head. The shoulders become firmly lodged in the pelvis, either owing to their size, or to their failure to rotate.

Treatment.—If the shoulders do not rapidly follow the head, firm pressure should be made upon the fundus. Avoid traction on the head, as it tends to make the shoulders become more firmly impacted. If they still do not advance, pass one hand into the vagina, and endeavour to get a finger into the posterior axilla; draw it down, and then in the same way pull upon the anterior axilla. If this fail, endeavour to bring down one or both arms. To do this, pass the whole hand into the vagina, and a couple of fingers upwards, along the posterior arm, until the elbow is reached. Then, by gentle pressure below the elbow, the forearm is made to flex, and the hand can be seized and drawn out. The anterior arm is then brought down in the same way. By this means, the width of the chest is diminished by twice the thickness of either shoulder. Traction can now be made upon both arms, and upon the head. If, in spite of this, the thorax do not follow, the hand should be introduced as far up into the uterus as possible, in order to determine if there be any pathological enlargement of the thorax or abdomen. If such a condition be present, it may be necessary to perform embryotomy.

CHAPTER XX.

CRANIOTOMY AND EMBRYOTOMY.

Craniotomy : Indications, Instruments, Conditions—Method : Perforation, Evacuation, Compression, Extraction—Perforation in the Case of a Face Presentation—In the Case of an After-coming Head—Embryotomy : Decapitation, Indications, Instrument, Method—Evisceration, Indications, Instrument, Method.

CRANIOTOMY.

By the term craniotomy, is meant any cutting operation performed upon the head of the fœtus, with the object of reducing its bulk.

Indications.—As craniotomy of necessity implies the death of the child, it must only be performed under conditions of absolute necessity. The indications for the operation are as follow :—

(1) If the child be dead, and if extraction of the undiminished head would be dangerous for the mother.

(2) If the child, in all probability, could not be extracted alive, and if such extraction would be dangerous for the mother.

(3) If the child be alive, and a relative indication for Cæsarean section or symphysiotomy exist (*v.* pp. 196, 200), but the mother refuse the operation (Winckel).

Instruments.—The instruments which are required, and which are best adapted for craniotomy, are :— a Simpson's perforator ; a Winter's modification of Auvard's combined cranioclast and cephalotribe (*v.* Fig. 23) ; and a large-size Bozeman's catheter.

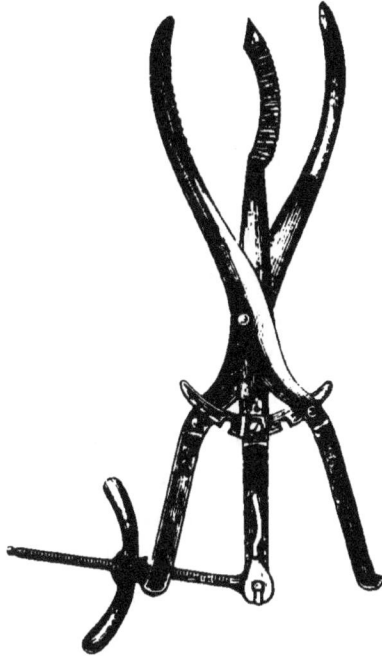

FIG. 23.—Winter's modification of Auvard's combined cranioclast and cephalotribe.

Conditions.—Certain conditions are necessary before the operation can be performed :—

(1) The pelvis must measure more than $2\frac{1}{2}$ inches in the *conjugata vera*. Extraction of even. a perforated head, through a pelvis which is smaller than this, is so dangerous that it should not be attempted.

(2) The os must be sufficiently dilated to permit the necessary manipulations.

(3) The head must be fixed, or be held fixed by an assistant.

Method.—The operation consists of four steps :—
(1) Perforation.
(2) Evacuation.
(3) Compression.
(4) Extraction.

Place the patient—previously anæsthetised—in the cross-bed position. Palpate the case carefully, and disinfect the parts thoroughly in the usual manner.

(1) *Perforation.*—Introduce as much of the hand as is necessary into the vagina, pass the fingers inside the os, and touch the presenting part. Slip the locked perforator upwards, under guard of the hand, and press it firmly through the centre of the presenting part, be it bone or suture. If this be done, there is less risk of the slipping of the instrument. Then release the catch which locks the perforator, and press the handles together; this separates the blades, so making a longitudinal cut in the calvarium. Withdraw the instrument partially, and turn it round through a right angle. Open the blades again, so making another cut at right angles to the former one.

(2) *Evacuation.*—Push the instrument, through the opening thus made, down to the base of the skull, and, moving it about, break up the brain thoroughly. Commence with the medulla oblongata in order to ensure the death of the child. Next introduce the Bozeman's catheter, and douche out the fragments of the brain.

(3) *Compression.*—Now take the combined cranio-

clast and cephalotribe. It consists of three blades ;—
a central or male blade, which resembles the male
blade of an ordinary Braun's cranioclast ; and two
outside blades, both of which lock into the central
blade. One of these outer blades locks with the central
blade so as to form a cranioclast; the other blade

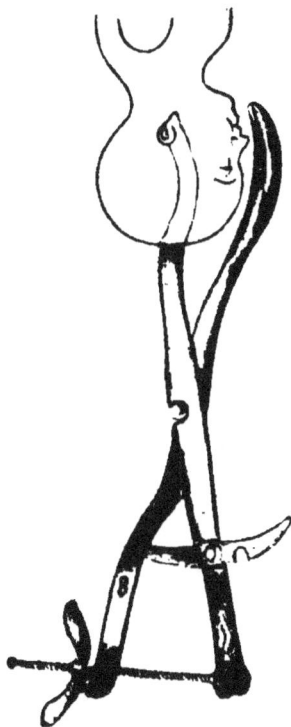

FIG. 24.—First step in the application of the combined cranioclast
and cephalotribe (modified from Dührssen).

completes the cephalotribe. The instrument is also
furnished with a strong screw, which can be adjusted so as to compress either external blade
against the central blade. To use it, introduce the
central blade into the interior of the cranium, and

then pass one of the external blades upwards, in such a manner that it lies over the face of the child (*v.* Fig. 24). Take care that the central blade is so turned that its convexity points towards the external blade, as otherwise it would not have so firm a grip

FIG. 25.—Second step in the application of the combined cranioclast and cephalotribe (modified from Dührssen).

upon the head. We have now a cranioclast upon the child's head, and, if the obstruction be not too great, the head can be delivered by it without using the other blade. A cranioclast acts by elongating the evacuated head, and so reducing all its diameters except the vertical.

In some cases it may be necessary to reduce the size of the head still further, and this can be accomplished by the aid of the third blade.

Having applied the cranioclast over the face of the child, and tightened the screw until the catch can be fastened, introduce the third blade, so that it lies at the opposite side of the head to the cranioclast (v. Fig. 25). Lock it, and apply the screw to it. Then tighten the screw until the handles come sufficiently close to enable the catch which holds the second blade to be fastened (v. Fig. 26). Always endeavour to pass the blades as far upwards, over the base of the skull, as possible.

The great advantage which Auvard's instrument possesses, is, that the head can be crushed without any fear of the cephalotribe slipping, as it is held firm by the previously applied cranioclast.

(4) *Extraction.*—Perforation should always be followed by extraction. In many cases, the contractions of the uterus would expel the perforated head without assistance, but it is not wise to allow this to occur. In the first place, decomposition proceeds very rapidly inside a perforated head, and the patient may thus become infected. In the next place, she has probably been allowed to remain undelivered as long as is safe, and therefore the uterus must now be emptied. Extraction is performed by means of the cranioclast or the combined instrument. In performing it, the head should be made to rotate in such a manner as to imitate, as nearly as possible, the normal mechanism of labour.

In the case of a *face presentation*, endeavour to introduce the perforator through one of the orbits, and, failing that, through the roof of the mouth.

In the case of the *after-coming head*, the operation of perforation is usually extremely difficult.

FIG. 26.—Final step in the application of the combined cranioclast and cephalotribe (modified from Dührssen).

Choice may be made between two sites for the introduction of the perforator. It may be introduced either into one of the lateral fontanelles, or into the occiput (Dührssen). If the former site be chosen, draw the body of the child forwards, and to one side ; by this means the lateral fontanelle is made to descend. If the latter site be chosen, draw the body forcibly backwards, introduce the fingers of the left hand between the symphysis and the occiput of the child, and perforate at the highest point which is protected by the fingers (Dührssen).

EMBRYOTOMY.

Embryotomy is the term applied to any operation intended to reduce the size or shape of the child's body. It includes decapitation and evisceration.

Decapitation.—By decapitation is meant the separation of the child's head from the body at the neck, so as to allow the child to be more easily extracted.

Indications.—It is indicated in cases of neglected shoulder presentation, when version is either impossible, or is contra-indicated, and in which the neck can be reached.

Instrument.—The best instrument for the purpose is Braun's blunt hook (*v.* Fig. 27). It performs the operation with ease, and with a minimum of danger for the mother.

Method.—To perform the operation of decapitation, place the patient, fully anæsthetised, in the cross-bed position. Introduce one hand into the vagina, and endeavour to encircle the neck with the fingers. Then pass the hook, under cover of the hand, upwards behind the symphysis ; turn it, so that it lies over the neck of the child ; and, finally, by a series of twisting movements tear through first the soft parts, and lastly the spinal column. Then draw

down the arms, and extract the trunk by traction
upon them.

The head is extracted last, and may cause some
trouble. The easiest method of extracting it is to
pass the hand into the uterus, and two fingers into
the mouth, and in this way to draw the head downwards, whilst an assistant at the same time applies

FIG. 27.—Braun's blunt hook for decapitation.

pressure to the fundus. It may be necessary to
perforate and crush the head, if the pelvis be contracted.

Evisceration.—Evisceration consists in making an
opening into the thorax or abdomen of the child,
and through it removing some of the viscera.

Indications.—It is indicated if the size of the
child's body obstruct delivery.

Instrument.—A Simpson's perforator, with which
to make the necessary opening, is all that is required.
A pair of sharp pointed scissors will answer equally
well.

Method.—The patient must be in the cross-bed
position as before. Introduce the perforator into
whatever portion of the trunk can be most easily

reached. Make an opening sufficiently large to allow the hand or the fingers, according as is necessary, to be introduced. Seize any of the larger viscera that present, and tear them away. In this manner the liver, lungs, and heart may be removed. When the size of the trunk is sufficiently reduced, pass the hand into the uterus, seize the feet, and extract the child as a breech presentation.

CHAPTER XXI.

CÆSAREAN SECTION—PORRO'S OPERATION— SYMPHYSIOTOMY.

Cæsarean Section: Indications, Method — Porro's Operation: Indications—Symphysiotomy: Indications, Operation, After-treatment, Dangers.

CÆSAREAN SECTION.

CÆSAREAN section is the term applied to the operation by which the abdomen of the mother is opened; the uterus incised; the child extracted through the opening thus made; the incision in the uterus stitched up; and the latter replaced in the abdomen. It thus differs from Porro's operation, in which, after the extraction of the child, the uterus is removed.

Indications.—The indications for Cæsarean section may be divided into two classes :—

A. Absolute indications,—in cases in which abdominal section is the only means by which the child can be delivered.

B. Relative indications,—in cases in which the child could also be delivered by some other operation; such as symphysiotomy, perforation, or induction of premature labour.

In the first group of cases Cæsarean section must be performed. In the second group, whether it be adopted, or not, depends upon several circumstances :—the period of pregnancy, the condition of the child, and the will of the mother. The indications are :—

A. (1) Absolute pelvic contraction, *i. e.* below two inches in the conjugata vera.

(2) Solid irremovable tumours blocking the pelvis ; as,—bony growths from the pelvic walls, myomata of the uterus, cancer of the cervix, or ovarian tumours.

(3) Extreme cicatrisation of any part of the vagina, sufficient to prevent its being dilated without the rupture of other organs (Winckel).

B. (1) Lesser degrees of pelvic contraction, *i. e.* pelves which measure from 2 to 3½ inches in the conjugata vera, in the case of a living child ; when it is impossible to perform symphysiotomy, or to induce premature labour.

(2) Lesser degree of obstruction from solid tumours as above, A. (2), if the child be alive.

Method.—In a book of this nature, it is possible to give only a very short outline of the operation. It should be performed, if possible, after the patient has come into labour, but before the membranes have ruptured. There is then the least danger of post-partum hæmorrhage owing to the presence of uterine contraction ; and the os will be sufficiently dilated to allow free escape of the lochia during the puerperium. For the performance of the operation, four assistants are advisable ;—one to give the anæsthetic, one to take charge of the infant after its extraction, and two to assist the operator. The steps of the operation are as follow :—

(1) Open the abdomen in the middle line ; the uterus appears in the wound.

(2) Open the uterus in the centre of the presenting part by means of an incision six inches in length, the edges of the abdominal wound being kept firmly pressed against the uterus by an assistant. If the placenta lie under the wound it must be torn through.

(3) The child is rapidly extracted by the head, the cord ligatured and divided.

(4) The uterus is lifted out of the abdomen ; and the assistant grasps the lower uterine segment tightly as far down as possible, in order to check hæmorrhage, while the operator passes a gauze towel round it.

(5) This towel is twisted tightly, and the assistant, letting go the uterus, holds the ends of it.

(6) The placenta, membranes, and blood-clots are removed from the uterus.

(7) The cervix is seen to be patulous, otherwise it must be dilated.

(8) The uterine incision is then stitched up, by alternate deep and superficial stitches. The deep sutures travel the entire thickness of the uterine wall, with the exception of the mucosa. The superficial stitches include the peritoneum, and a little of the muscular coat.

(9) The peritoneal cavity is cleansed, and the abdominal wound stitched up by alternate deep and superficial sutures.

The after-treatment of the case resembles that of any abdominal section. The abdominal sutures are removed on the eighth day ; and, if all goes well,

the patient is allowed out of bed at the end of three
weeks. An abdominal binder must be worn for a
year.

PORRO'S OPERATION.

Porro's operation differs from Cæsarean section,
in that the uterus is removed supra-vaginally after
the extraction of the child.

Indications.—Porro's operation is indicated instead
of Cæsarean section in the following cases :—

(1) If the uterus be defectively developed.

(2) If the uterus be the subject of some incurable
disease, as cancer or myomata.

(3) If the patient suffer from osteomalacia.

(4) If we have reason to believe that the uterus
has been infected with septic organisms during
labour.

(5) It is also indicated in bad cases of concealed
accidental hæmorrhage (*v.* page 204).

The operation is the same as a supra-vaginal
amputation of the uterus, with extra-peritoneal
treatment of the stump. In some cases it may be
better to perform pan-hysterectomy, and to remove
the entire uterus.

SYMPHYSIOTOMY.

The operation of symphysiotomy consists in
dividing the ligaments of the symphysis pubis,
and so allowing the separation of the innominate
bones. As a result, all the diameters of the pelvis
are increased.

Indications.—Symphysiotomy is indicated in contracted pelvis when the conjugate diameter measures from $2\frac{3}{4}$ to $3\frac{1}{2}$ inches in length, if the induction of premature labour be out of the question. By dividing the symphysis an average separation of the pubic bones of about $2\frac{3}{5}$ inches occurs, and this yields an increase in the conjugata vera of $\frac{3}{5}$ inch.

As much as $3\frac{3}{5}$ inches separation of the pubic bones has been obtained with safety, yielding an increase in the conjugata vera of a little over $\frac{4}{5}$ inch.

Operation.—Four assistants are required;—one to assist the operator, one to give an anæsthetic, and two to control the separation of the innominate bones, which tend to spring suddenly apart the moment the symphysis is divided.

The end of the first stage is the best time for the operation. The steps of it are as follow :—

(1) An incision is made through the skin and subjacent tissues, starting about an inch and a half above the symphysis, and ending at the clitoris.

(2) An opening is made through the aponeurosis of the recti, sufficiently large to permit of the finger being passed behind the symphysis.

(3) The bladder and other retro-pubic structures are separated from the back of the symphysis, and from the pubic bones, for a distance of two inches at either side. This is a most important part of the operation, as it avoids laceration of the urethra, &c., when the bones separate.

(4) The pubic ligaments are divided, from behind forwards, and from above downwards, the

assistants taking every precaution to prevent the bones from springing asunder suddenly.

(5) The child is extracted by forceps or version.

(6) A piece of iodoform gauze is placed behind the separated bones, in order to prevent any of the soft parts from becoming nipped between them.

(7) The skin and subjacent parts are brought together by deep silk sutures, which go right down to the bone. As they are tied, the assistants press the pelvic bones again into contact.

(8) The wound is dressed in the usual manner, and a tight binder pinned round the pelvis in order to keep the bones in contact.

After-treatment.—The patient must lie upon her back on a hard bed for three weeks. The catheter must be passed regularly, and the binder changed as often as it becomes soiled. The patient may usually be allowed out of bed in five weeks, but a pelvic binder must be worn for a year after the operation.

Dangers.—The dangers of the operation are :—

(1) Rupture of the urethra.

(2) Hæmorrhage from laceration about the clitoris.

(3) Rupture of the sacro-iliac articulations.

(4) Failure of the joint to reunite.

CHAPTER XXII.

ANTE-PARTUM HÆMORRHAGE.

Varieties — Accidental Hæmorrhage : Ætiology, Varieties of
Accidental Hæmorrhage—Concealed Accidental Hæmorrhage ;
Symptoms, Treatment—External Accidental Hæmorrhage :
Symptoms, Diagnosis, Treatment by plugging the Vagina,
Other Modes of Treatment — Unavoidable Hæmorrhage :
Ætiology, Varieties, Symptoms, Diagnosis, Treatment by
Braxton Hicks's Method, Other Modes of Treatment—Fœtal
Mortality in Ante-partum Hæmorrhage.

ANTE-PARTUM hæmorrhage occurs in the later
months of pregnancy as two distinct varieties :—

(1) Accidental hæmorrhage.

(2) Unavoidable hæmorrhage, or hæmorrhage
due to placenta prævia.

ACCIDENTAL HÆMORRHAGE.

Accidental hæmorrhage is the term applied to
the hæmorrhage which results from the detachment
of a normally situated placenta.

Ætiology.—Accidental hæmorrhage is due to
almost the same factors as abortion. It is primarily
due to an endometritis, which may exist *per se*, or
may be caused by a variety of conditions, of which
perhaps the most constant is nephritis. If a patient
suffer from endometritis, any sudden movement

may cause the detachment of the placenta, and so hæmorrhage may commence.

Varieties.—There are two varieties of accidental hæmorrhage :—(1) concealed, (2) external. These differ from one another in the conditions which permit of their occurrence, and in the treatment which is suitable for them.

Concealed Accidental Hæmorrhage.—This is, perhaps, with the exception of acute sepsis, the most serious accident to which pregnant women are liable. It is, happily, very rare. In this condition, the blood which is poured out from behind the detached placenta is stored up in the uterus, which dilates in order to make space for it. The patient can thus bleed to death, although no blood escape into the vagina ; but it is only a uterus which is the subject of advanced metritis, which will dilate to this extent before the blood-pressure. It is an obvious fact that blood can escape from a ruptured vessel, into any cavity, only so long as the pressure inside the cavity is less than the blood-pressure. If the escaping blood flow out of the cavity as quickly as it flows in, then an indefinite amount can be lost. If the blood cannot escape, then it must cease flowing as soon as the cavity is full. There is no room for any considerable quantity of blood to escape into a healthy uterus, occupied by an unruptured ovum. If a vessel rupture in such a case, and no blood escape through the os, the pressure in the uterus would rapidly become greater than the blood-pressure, and the hæmorrhage would cease. If, on the other hand, the uterus be unhealthy, and dilate before the blood-pressure, then the amount of the hæmorrhage is only limited by the dilatability of the uterus.

This is an important fact to grasp thoroughly, as it shows :—

(1) How concealed hæmorrhage occurs.

(2) The method by which external accidental hæmorrhage in a healthy uterus may be checked.

(3) How useless it would be to adopt this same method in concealed accidental hæmorrhage, *i. e.* in the case of a diseased uterus.

Symptoms.—The symptoms of concealed accidental hæmorrhage are those common to any form of internal hæmorrhage. Collapse, falling temperature, weak and rapid pulse, severe abdominal pain, anæmic appearance,—all occur in proportion to the amount of blood which the patient is losing. At the same time the uterus increases in size, becomes tender to the touch, and there is an increasing difficulty in feeling the fœtus.

Treatment.—The only treatment which is of any avail in these cases is *accouchement forcé*, or Porro's operation.

Accouchement forcé consists in rapidly dilating or incising the cervix, turning the presentation into a footling presentation, if it be not one already, and then extracting the child by applying traction to the leg. If the hæmorrhage continue, the uterus must be immediately plugged.

The choice between the two lines of treatment depends largely upon the skill and experience of the practitioner, and upon the circumstances under which the operation has to be performed. In a hospital, where all the requisites for abdominal section are present, Porro's operation is the better line of treatment.

External Accidental Hæmorrhage.—This is also a very serious complication of pregnancy, although the prognosis is not nearly so bad as in concealed hæmorrhage. The blood escapes from the uterus as rapidly as it flows out of the ruptured vessels, and so the hæmorrhage is at once apparent.

Symptoms.—The escape of blood is the first symptom, accompanied by a varying degree of pain. If the case be untreated, the usual symptoms of hæmorrhage follow.

Diagnosis.—The diagnosis has to be made from hæmorrhage due to placenta prævia, and, as a rule, it is easy to do so. Examine the patient vaginally; if the placenta can be felt through the os, or through the lateral fornices, it is a case of placenta prævia. If the placenta cannot be felt, it may possibly be a case of lateral placenta prævia, but it is to be treated as if it were a case of accidental hæmorrhage. The condition may also be diagnosed by abdominal palpation. If the head be found to be fixed in the brim, it is almost certainly not a case of placenta prævia.

Treatment.—The treatment, and the gravity of the case, depend entirely upon whether the patient be, or be not, in labour. If she be in labour, the danger of the condition is greatly diminished, and the treatment is simple. If she be not in labour, the reverse is the case. I pointed out above, the conditions under which concealed hæmorrhage occurs; and I also showed that there was no room for blood to be stored up in a healthy uterus if the ovum were intact. It is on this fact that the treatment which I am about to describe depends.

If we prevent the blood, which is escaping from behind the placenta, from leaving the uterus, the pressure inside the latter will rapidly become greater than the blood pressure, and, as a result, the hæmorrhage will cease. How, then, can the escape of blood from the cervix be prevented? By plugging the vagina tightly. This will check the hæmorrhage, and at the same time bring on labour—the two results which we most wish for under the circumstances. At the same time labour is brought on gently, without causing any aggravation of the shock from which the patient is usually suffering. On the contrary, she is given ample time to rally from the collapse which the hæmorrhage caused, before the uterus empties itself.

To perform the operation of plugging, place the patient in the cross-bed position, wash and douche her thoroughly. Anæsthesia is not necessary. Then pass a posterior speculum, and with strips of iodoform gauze, soaked in lysol solution, plug tightly round the cervix. The rest of the vagina is then plugged, as firmly as possible, with balls of absorbent wool about the size of a large walnut, also soaked in lysol solution. The plugging is continued until the vagina is as full as it will hold. The patient is then put back to bed, and a tight abdominal binder applied. The wadding which is used should have been previously immersed for five or ten minutes in boiling water. The plug is left in until strong labour pains ensue, this usually occurring in from three to four hours. In some cases the onset of labour may be slower than this, and then the plug must be removed after twelve hours, for fear of decomposition. If the hæmorrhage come on again she must be replugged, but this is usually unnecessary. The success of this treatment depends upon two points :—

(1) The ovum must be intact.

(2) The plug must be applied tightly.

If the patient be in labour at the time the hæmorrhage commences, it is not a difficult matter to check the latter. If the pains be strong and the membranes intact, rupture them. This enables the head to advance without at the same time pulling upon the placenta, and so detaching more of it. Also, owing to the escape of the liquor amnii, the uterus is enabled to contract down upon the child, and thus to diminish the size of the placental site. If the hæmorrhage still continue, the child may be delivered by version, followed by extraction ; or, if the head be fixed and the os dilated, by the application of forceps.

Other Modes of Treatment.—Other modes of treatment, which are recommended, are :—

(1) Rupture of the membranes in every case. This would be good treatment if we could be certain that the uterus would contract down upon the child. But we cannot be certain of this, unless the patient have strong labour pains. It should, therefore, be reserved for these cases.

(2) *Accouchement forcé.* This is exceedingly bad treatment. If the patient have lost much blood, she is in danger of dying of shock. Any intra-uterine manipulations increase this shock, particularly if they be followed by the forcible

extraction of the child. The only point in favour of this treatment is that it gives a lower infantile mortality. This point will be referred to again (*v.* page 213).

PLACENTA PRÆVIA.

By placenta prævia is meant the implantation of the placenta so near the internal os that a portion of it is torn off, during the formation of the lower uterine segment (Winckel).

Ætiology.—The cause of placenta prævia is not very definitely known. Three theories as to its occurrence are worthy of consideration :—

(1) That, owing to an antecedent endometritis, the cavity of the uterus is enlarged, and its walls are not in as close contact, one with the other, as is normal. As the result of this, when the ovum enters the uterus, instead of being detained in the neighbourhood of the Fallopian tube, it drops down, and becomes implanted in the lower uterine segment.

(2) That the placenta is formed out of the chorionic villi which cover the lower segment of the ovum instead of out of those that are attached near the fundus of the uterus.

(3) That a placenta prævia is merely a normally situated placenta, which, owing to its size, has invaded the lower uterine segment.

Whatever may be the actual cause of the condition, there is little doubt that endometritis plays an important part in it.

Varieties.—Three varieties of placenta prævia are described :—

(1) *Placenta prævia centralis;* in which the placenta covers the entire undilated internal os.

(2) *Placenta prævia marginalis;* in which the placenta comes down to the edge of the undilated internal os.

(3) *Placenta prævia lateralis;* in which a portion of the placenta lies in the lower uterine segment, but does not descend so far as the undilated internal os (*v.* Fig. 28).

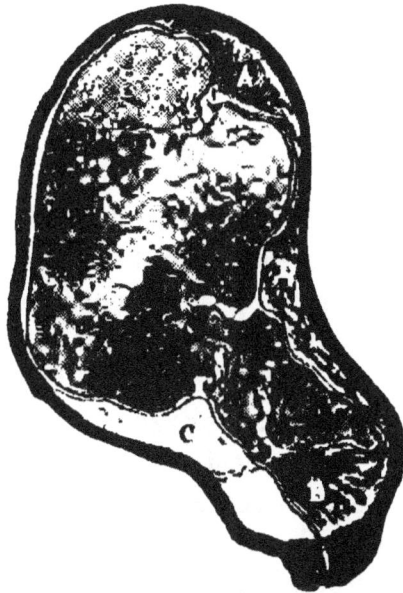

FIG. 28.—Diagram representing the various situations of the placenta. A. normal situation; B. Placenta prævia centralis; C. Placenta prævia marginalis; D. Placenta prævia lateralis. (Modified from 'The Norris Text-book of Obstetrics.')

Symptoms.—The chief symptom is hæmorrhage, occurring any time after the commencement of the seventh month. If the hæmorrhage continue untreated there will be the usual symptoms of collapse.

Diagnosis.—The diagnosis is made by examining the patient vaginally. If the placenta be felt through the os or through the lateral fornix, it is a case of placenta prævia. If the placenta cannot be felt, then it is either a case of accidental hæmorrhage or of hæmorrhage due to a placenta prævia lateralis; in either case the treatment to be adopted is that of accidental hæmorrhage. In a favorable subject the occurrence of placenta prævia may be diagnosed by abdominal palpation. The points to be looked for are :—the high situation of the presenting part, and an increased sense of resistance, and an increased difficulty in feeling the fœtal parts, over any one portion of the lower uterine segment.

Treatment.—As soon as the diagnosis of placenta prævia is made the case should be treated. As in accidental hæmorrhage, the treatment depends on whether the patient be, or be not, in labour. Usually she is not in labour at the onset of the hæmorrhage, and in this case the prognosis is considerably more serious. The best treatment then is that recommended by Braxton Hicks. It consists in turning the fœtus by bipolar version into a breech presentation, rupturing the membranes, drawing down a foot (*v.* page 177), and leaving the rest of the delivery to nature. If it be a case of central placenta prævia, the fingers must be pushed directly upwards through the placenta in their attempt to seize the foot. This treatment both checks the hæmorrhage, by the pressure of the breech or back of the child against the placenta, and brings

on labour. A piece of gauze should be tied to
the foot; and, if further hæmorrhage occur, light
traction on the gauze will check it, by drawing
down more of the breech. This treatment requires
two conditions to be present :—
  (1) The membranes must be unruptured.
  (2) The os must be large enough to admit
      at least two fingers.
  The first condition is practically always present,
unless, indeed, an ignorant attendant have ruptured
the membranes. The second condition is present
in more than 99 per cent. of all cases of placenta
prævia in which the patient is bleeding. In the
rare instance in which it is not present, plug the
vagina and leave the plug in for a few hours.
The os will then be found sufficiently dilated to
allow version to be performed.
  If the patient be getting strong labour pains
when the hæmorrhage commences, rupture of the
membranes is often sufficient to check the bleeding. Rupture of the membranes acts in the
same way, in these cases, as it does in accidental
hæmorrhage in the presence of strong labour pains;
—viz. it allows the head to advance without detaching more of the placenta, and it diminishes
the size of the placental site (v. page 207). If
the hæmorrhage still continue, the child may be
delivered by forceps, if the head be fixed and the
os dilated. If these conditions be not present, the
child may be turned by internal version and the
rest of the delivery left to nature.
  Other Modes of Treatment.—Other modes of
treatment recommended are :—

(1) *Accouchement forcé.* The same objection applies to the adoption of this treatment in the case of placenta prævia, as does to its adoption in accidental hæmorrhage (*v.* page 207).

(2) Plugging the vagina. There is a very considerable risk of sepsis in any case of plugging, more particularly in placenta prævia, owing to the low situation of the placenta. It should not be resorted to unless it be absolutely necessary.

(3) Partial detachment of the placenta—Barnes' method. The patient runs a greater risk of sepsis, if this method be adopted, than if version be adopted. In performing it the fingers of necessity come into very close contact with the uterine sinuses, and bacteria may be introduced if the strictest asepsis be not adopted.

Complications.—Patients suffering from placenta prævia are more liable to post-partum hæmorrhage and to sepsis than are others. The former frequently occurs, owing to the fact that the lower uterine segment does not contract as firmly as the fundus, consequently the uterine sinuses may be only partially obliterated. If rapid dilatation of the cervix and extraction be the treatment adopted, deep lacerations of the cervix are almost certain to occur. The cervix tears very much more easily in cases of placenta prævia than in cases of normal insertion of the placenta. It thus frequently happens, that, while we think that the os is dilating under the pressure of our fingers, it is really tearing. Again, a laceration of the cervix which would be trivial in the case of a normally situated placenta, may cause grave trouble in placenta prævia, owing to the large vessels in the neighbourhood of the cervix which supply the placenta. There is also more risk of sepsis in these cases. The placental site lies so near the vagina, that, if any putrefaction occur in the latter, the former is almost certain to become infected.

It is well said, that, in any case of ante-partum hæmorrhage, the life of the child must be considered as antagonistic to the life of the mother. Any treatment which yields the smallest fœtal mortality will give the largest maternal mortality, and *vice versâ*. Thus, in both accidental and unavoidable hæmorrhage, *accouchement forcé* will save the greatest number of children; but it will lose from six to twelve times as many mothers as will the treatment by plugging in accidental hæmorrhage, or by bi-polar version in placenta prævia. Even if the child be brought into the world alive, in either of these conditions, it is most frequently premature and weak from its previous semi-asphyxia. As a result, it almost invariably dies within the first month. Under these circumstances, the life of the mother should surely not be risked, by adopting a treatment which is avowedly more dangerous for her, merely because it affords a slightly improved chance of saving the child.

CHAPTER XXIII.

POST-PARTUM HÆMORRHAGE.

Varieties — Traumatic Post-partum Hæmorrhage: Diagnosis, Treatment—Atonic Post-partum Hæmorrhage: Mechanism by which Hæmorrhage is checked after the Detachment of the Placenta, Ætiology, Treatment, Plugging with Iodoform Gauze, Injection of Perchloride of Iron — Treatment of Collapse due to Hæmorrhage : Transfusion of Saline Solution.

THE term post-partum hæmorrhage is, for convenience sake, applied to any hæmorrhage which occurs after the birth of the child, irrespective of the fact that parturition may, or may not, be complete.

Varieties.—There are two distinct varieties of post-partum hæmorrhage :—

(1) Traumatic.

(2) Atonic.

Traumatic post-partum hæmorrhage is the term applied to hæmorrhage resulting from a laceration of some part of the genital canal. There are two situations in which a laceration is likely to cause hæmorrhage ; these are,—in the neighbourhood of the clitoris, or in the neighbourhood of the cervix.

Diagnosis.—The diagnosis has to be made from atonic post-partum hæmorrhage, that is hæmorrhage due to the failure of the uterus to contract. This is easily accomplished by placing the hand on the fun-

dus ; if it be firm and well contracted it is obvious
that the hæmorrhage cannot be due to the failure
of the uterus to contract, *i. e.* it must be traumatic
hæmorrhage. In some cases, the diagnosis may not
be made until we have commenced to douche the
uterus. If a Bozeman's return catheter be used for
this purpose, it at once distinguishes between the
two conditions. In the case of hæmorrhage from the
interior of the uterus, the solution which flows back
through the return pipe of the catheter will be
blood-stained. If the hæmorrhage be from a laceration outside the uterus, the solution in the return
pipe will be colourless, whilst at the same time blood
is seen to flow from the vagina or vulva.

Treatment. — If traumatic hæmorrhage be sus-
pected, examine the region of the clitoris. If
there be a laceration which is bleeding, it must be
stitched. Pass a silk ligature by means of a small
curved needle deeply under one end of it, going
right down to the bone, and a second ligature at
the other end. Tie them tightly, and the hæmorrhage at once ceases. The ligatures must be
removed in six or seven days. If, on inspection, no
laceration can be detected about the clitoris, we
must examine the cervix. To do this draw it down
with a bullet forceps, if there be one to hand, and
examine it carefully for a laceration or bleeding
vessel. A laceration must be stitched up, a bleeding vessel must be tied. If no bullet forceps be to
hand to draw down the cervix, an equivalent can
be extemporised by means of a ligature. Thread a
curved needle with a long ligature, and pass the
needle held in a needle-holder up to the cervix under

cover of the fingers in the vagina. Then pass the needle through the first part of the cervix that comes to hand, and draw it through, leaving the ligature *in situ*. By means of it the cervix can be drawn down and exposed. Pressure upon the fundus, by causing descent of the uterus, is also of great assistance.

Atonic post-partum hæmorrhage is due to failure of the uterus to contract and retract. Before entering into the causes of this condition, it is well to understand what it is that normally prevents the occurrence of hæmorrhage after the detachment of the placenta. This end is brought about by three factors:—

(1) The contractions of the muscular coat of the uterus.

(2) The retraction of the muscle-fibres of the uterus.

(3) The clotting which occurs in the mouths of the vessels.

I have explained the difference between contraction and retraction before (*v.* page 56), and also that the former is intermittent, the latter continuous. The latter is therefore the more important. We may, then, attribute the non-occurrence of post-partum hæmorrhage to the permanent retraction of the uterine fibres, which takes place after the child is expelled. The contractions of the uterus will check all hæmorrhage during the period of contraction, but retraction, once established thoroughly, prevents the occurrence of any further hæmorrhage. The third means by which hæmorrhage is checked is of little or no importance, *i.e.* the clotting of blood in the open mouths of the vessels. It is probably the result of the checking of the hæmorrhage, and not the cause of it.

Ætiology.—We can now understand the causes of atonic post-partum hæmorrhage. Considered generally, they are anything which interferes with the due retraction of the uterine muscle-fibres. They are as follow:—

(1) Retained placental fragments, membranes, or blood-clots, *i. e.* bad management of the third stage of labour.

(2) Tumours of the uterus, as myomata.

(3) Over-distension of the uterus, as in hydramnios or twins.

(4) Metritis.

(5) Protracted labour.

(6) Precipitate labour.

(7) Previous hæmorrhages which have weakened the patient. Similarly,—

(8) Any weakening disease.

(9) Uterine inertia, starting in the first and second stages of labour.

(10) Placenta prævia, in which part of the placenta is situated below the contractile portion of the uterus.

Treatment.—The most important point, in the successful treatment of post-partum hæmorrhage, is to have a definite plan of action laid out,—a plan which commences with the mildest measures, and goes gradually on to more serious measures if necessary. I shall describe such a plan, giving it in the order that should be adopted, and supposing that the failure of each measure in turn requires the adoption of the subsequent one.

(1) Ascertain whether the placenta be in the uterus or in the vagina. If it be in the uterus, stimulate the fundus to contract by gentle friction. If the hæmor-

rhage still continue, or if the placenta be in the vagina at the beginning,

(2) Endeavour to express it after "the manner of Credé"; if that cannot be done, remove it manually (v. page 228).

(3) Stimulate the fundus to contract by friction, and administer ergot. Up to three drachms of the liquid extract (B. P.) may be given; but far more certain and reliable are the hypodermic tabloids of citrate of ergotinin; up to $\frac{1}{25}$ of a grain of this preparation may be administered hypodermically.

(4) Wash the patient externally, and then douche the vagina with creolin solution at a temperature of 110°—115° F., having first passed a catheter.

(5) Douche the uterus thoroughly with the same solution.

(6) Compress the fundus firmly between the fingers of one hand in the anterior fornix, and the other hand upon the abdominal wall, thus squeezing out any clots that may be retained. Then repeat the intra-uterine douche.

(7) Introduce the hand into the uterus, and remove any fragments of placenta or membranes and all clots that may be in it. Then repeat the intra-uterine douche.

(8) Now choice must be made between two lines of treatment, as the adoption of one excludes the other. These are,—either to plug the uterus with iodoform gauze, or to inject perchloride of iron.

The uterus is plugged with iodoform gauze in the following manner. Seize the anterior lip with a bullet forceps, and pass a posterior speculum if one can be obtained. Then pass a long strip of iodoform gauze up to the fundus, by means of a plugging forceps, or with the end of the Bozeman's catheter. The rest of the strip is then pushed into the uterus, taking care to pass it as far up towards the fundus as possible. If another strip have to be used, it must be knotted to the first strip in order to facilitate its extraction. It must be remembered, that it is not the large cavity of a dilated uterus which we have to plug, but only the small cavity of a contracted uterus; because, on the introduction of a small quantity of gauze, the uterus, which before was flaccid and relaxed, quickly contracts firmly upon the foreign body. A tight abdominal binder is then applied in order to control the uterus from above, and the patient is put back to bed. The gauze is removed in twenty-four hours, and the uterus thoroughly douched out.

The use of perchloride of iron in these cases was introduced by Barnes. He recommends that a few ounces of the Liq. Ferri Perchlor. (B. P.) be injected into the uterus, from which all clots have been previously removed. An easier method is to add Liq. Ferri Perchlor. fort. (B. P.) to the douche until a light sherry-coloured fluid be produced. The uterus is douched out thoroughly with this, and then with the hot creolin solution. Barnes claims that iron acts in three ways :—

(1) It coagulates the blood in the mouths of the vessels.

(2) It constringes the tissues round the vessels, and thus compresses them.

(3) It provokes some contraction of the muscular wall of the uterus.

The objection to perchloride of iron is that it

always causes a certain amount of superficial necrosis of the uterine wall. If saprophytic germs gain access, they have then a suitable nidus in which to lodge. Again, if iron fail to check the hæmorrhage, plugging is impossible, owing to the manner in which the tissues have become constringed. The objection to gauze is that it may not be sterile. If we can be sure that it is sterile, then plugging is much the better line of treatment. If iron be injected, the uterus must be thoroughly douched out the next day.

When a patient is attacked by any kind of hæmorrhage, there are two chief indications for treatment:—

    (1) The hæmorrhage must be checked.
    (2) The collapse which threatens to follow the hæmorrhage must be staved off.

I have described how the hæmorrhage may be checked; I shall now consider the treatment of the collapse. When a patient loses a quantity of blood, death threatens. This occurs not because there is an insufficient quantity of blood in the body, but because the blood-vessels have not had time to contract to suit their capacity to the diminished quantity of blood. Blood consequently does not return to the heart in sufficient quantities; the latter has not sufficient fluid to contract upon; as a result, its contractions become more and more feeble, and an insufficiency of blood is sent to the brain. In consequence of this, feeble stimuli are transmitted to the heart, which fails still more, a vicious circle being thus established. Reasoning from this, we see that it is necessary to turn our treatment in three directions :—

    (1) The heart must be stimulated.

(2) The diminished quantity of blood must be limited, as far as possible, to the important organs of the body, *i. e.* the brain and the viscera.

(3) The amount of fluid in the blood-vessels must be increased.

We can stimulate the patient by giving alcohol by the mouth ; by the hypodermic injection of ether, strychnine, or brandy ; and, by the use of hot fomentations over the heart. We can keep the blood in the important organs, first, by placing blocks beneath the foot of the bed, and thus making the patient's head the most dependent part of her body ; subsequently, by bandaging tightly the arms and legs, and thus preventing blood from being wasted upon them. We can increase the quantity of fluid in the blood-vessels by giving plenty of fluid by the mouth ; by administering enemata of salt and water ; and by infusing saline solution directly into a vein, or into the connective tissues of the breast, axilla, or buttock.

There are no special points in the above treatment which call for description, except the method of infusing saline solution into a vein. The apparatus required for this purpose is very simple. It consists of :—a small metal funnel which holds about three ounces ; a rubber tube of about three feet in length ; and a small silver cannula (*v.* Fig. 29). The solution used is made by adding a teaspoonful of salt to a pint of water. If possible it must be sterilised by boiling, and must be administered at a temperature of 100° F. The operation itself is as follows : —Tie a bandage round the upper arm sufficiently tightly to compress the veins, but not the arteries.

By this means the veins are made to stand out, and
a suitable one can be selected. Expose the vein by
means of an incision about an inch long made
directly over it; isolate a very small portion,
and then slip two ligatures beneath it. The
distal ligature is tied to prevent hæmorrhage ; the
vein is opened by a longitudinal incision sufficiently
long to admit the cannula ; and the latter is introduced. Next tie with a single turn the proximal
ligature, in such a manner that the vein is compressed

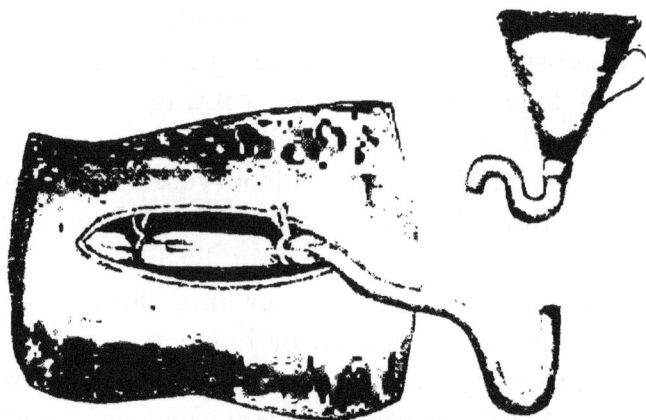

FIG. 29.—Apparatus, and method of inserting cannula, for intra-
venous infusion of saline solution (diagrammatic).

against the cannula, in order to prevent the escape of
the solution ; and remove the bandage which was
compressing the arm (v. Fig. 29) Before the cannula is introduced the entire apparatus must be filled
with the saline solution, the escape of it being con-
trolled by pressure upon the tube. The fluid is now
allowed to flow, an assistant taking care that the
funnel is always filled with solution. As many as
five, six, or even more, pints may be injected in

severe cases. When sufficient has been injected the cannula is removed, the second ligature tied tightly, the vein cut across, and the skin wound stitched up with a continuous suture.

Let me conclude this chapter with a well-known remark : " Your patient should not be allowed to die of post-partum hæmorrhage."

CHAPTER XXIV.

PRECIPITATE LABOUR—UTERINE INERTIA—RETAINED PLACENTA.

Precipitate Labour: Treatment—Uterine Inertia: Varieties—
Primary Uterine Inertia: Ætiology, Symptoms, Treatment—
Secondary Uterine Inertia; Ætiology, Symptoms, Treatment
—Retained Placenta: Ætiology, Treatment.

PRECIPITATE LABOUR.

Precipitate labour occurs, when the contractions
of the uterus are considerably stronger than are
necessary to overcome the resistance of the soft
parts of the mother. As a result, the child is driven
rapidly through the pelvis, and is born when, perhaps,
the mother is not in a suitable position. In consequence of this, the umbilical cord may be torn, the
placenta may be detached prematurely, or even the
death of the child may result.

Treatment.—If we know that a patient is subject
to precipitate labour, she should be placed in bed
immediately labour commences, and not be allowed to
leave it. By this means accidents will be prevented.

UTERINE INERTIA.

By uterine inertia, we mean that the contractions of
the uterus are feeble, so that they either fail altogether

to expel the child, or only succeed after a long time.

Varieties.—Uterine inertia occurs in two distinct forms :—(1) primary inertia ; (2) secondary inertia. These are so distinct one from the other that they must be considered separately.

Primary Uterine Inertia. — In this condition, the contractions of the uterus are more feeble than normal, from the very commencement of labour. The uterus never contracts strongly.

Ætiology.—The causes of primary uterine inertia lie in the uterus itself, or in its contents. They are :—

(1) Weak muscular development.

(2) Faulty shape, as uterus bicornis.

(3) Metritis.

(4) Over-distension, as by hydramnios or twins.

(5) Tumours, as myomata.

Symptoms.—The os dilates slowly ; there is only slight distension of the bag of membranes during a pain ; no caput succedaneum forms upon the child's head ; and the hardening of the uterus during a pain is almost imperceptible. If the head lie in the pelvis for a very long time, the patient becomes feverish and restless, and sloughing of the vaginal walls or cervix may occur. The third stage is usually cha-racterised by the slow expulsion of the placenta, or by its non-expulsion, and perhaps by the occurrence of atonic post-partum hæmorrhage.

Treatment.—In primary uterine inertia, the uterus is obviously not sufficiently strong to expel the child, and the indication is to assist it. This may be done by massage of its walls, followed by the expression

15

of the fœtus if possible (Kristeller's method). If such treatment be not successful, forceps must be applied, as soon as the necessary conditions are fulfilled (*v.* page 169).

Secondary Uterine Inertia.—In this condition, the contractions of the uterus may have been of normal intensity at the commencement of labour, but have gradually diminished in strength as labour proceeds.

Ætiology.—A lesser degree of the same pathological conditions of the uterus that caused primary inertia, may also cause secondary inertia. To them may be added the following :—

(1) Distension of the bladder or rectum.
(2) Contracted pelvis.
(3) Large head.
(4) Pendulous abdomen.
(5) Weakness or collapse of the patient.
(6) Rigid soft parts.

Symptoms.—The symptoms are similar to those of primary inertia.

Treatment.—If any obstruction be present, such as a full bladder or rectum, remove it. Correct any obliquity of the uterus by applying an abdominal binder. Give the patient an opiate, as 20 to 30 minims of Tinct. Opii ; this will cause her to sleep, and when she awakes, she will be refreshed, and the pains may return. If she still fail to deliver herself, forceps may be applied, if the necessary conditions be present.

RETAINED PLACENTA.

I have already said that if the uterus fail to expel the placenta, it must be made to do so. It may expel the placenta immediately after the birth of the child, or it may not expel it for an hour. If the uterus have not expelled the placenta spontaneously within this period, steps must be taken to expel it artificially.

Ætiology.—There are four chief causes of retained placenta :—

(1) Uterine inertia.

(2) Morbidly dense adhesions between the placenta and the uterus.

(3) A *placenta membranacea*.

(4) Hour-glass contraction of the uterus.

Uterine inertia causes the placenta to be retained, owing to absolute failure of the uterus to contract. Morbid adhesions between the placenta and the uterus are the result of endometritis ; they may be so dense that it is almost impossible to remove the placenta. A *placenta membranacea* is the term applied to a large, thin, membrane-like placenta. When the uterus contracts, it crumples up such a placenta in its interior instead of detaching it. Hour-glass contraction is a very rare condition. The uterus contracts circularly below the placenta, while the fundus remains uncontracted. It is practically always due to bad management of the third stage. The attendant massages the lower uterine segment instead of the fundus ; so causing the former to contract and prevent the expulsion of the placenta.

Treatment.—If the retention be due to hour-glass contraction, cease massaging the uterus ; the contraction will then probably pass off, and the placenta will be expelled. If it be not expelled, or if there be hæmorrhage and we cannot wait, introduce the fingers into the uterus in the shape of a cone, push them gently and slowly through the obstruction, and remove the placenta. Care must be taken to do this slowly and without force, or the uterus may be ruptured.

In retention of the placenta, due to other causes, massage the fundus, and attempt to express the placenta by the so-called " method of Credé " (v. page 78). If this fail it must be removed manually. This operation, which used to be considered one of the most dangerous in midwifery, owing to the risk of sepsis, is now performed with perfect safety if the usual aseptic precautions be used. It is performed as follows :—Place the patient in the cross-bed position, wash her thoroughly externally, and empty the bladder. Introduce the hand into the uterus, taking care to keep outside the membranes. Feel for the edge of the placenta, and with the tips of the fingers separate it from the uterus with a sawing motion. Endeavour to detach it all in one piece. When it is completely detached, grasp it in the hand and remove it. Introduce the hand again into the uterus to feel if any portions have been left behind. As soon as all the fragments have been removed douche the uterus thoroughly, and put the patient back to bed. Never give an anæsthetic in these cases if it can be avoided. If it must be given, let the patient be fully under its influence, before the hand be introduced into the uterus.

CHAPTER XXV.

LACERATIONS OF THE GENITAL TRACT.

Ruptured Uterus : Ætiology, Symptoms, Treatment, Dangers
—Laceration of the Cervix—Laceration of the Perinæum :
Dangers, Treatment.

RUPTURE OF THE UTERUS.

RUPTURE of the uterus may occur at any stage of
labour. It is a rare accident, but is perhaps not quite

FIG. 30.—Rupture of thinned-
out lower uterine segment.
(Spiegelberg.)

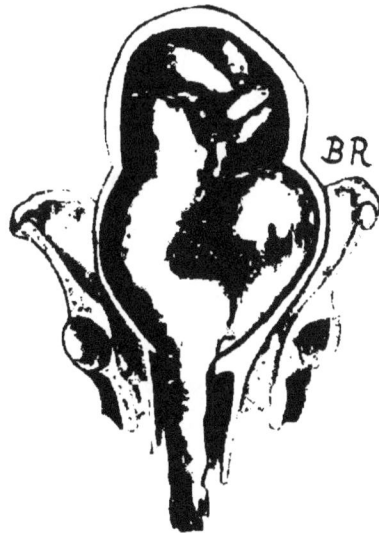

FIG. 31.—Thinning of lower
uterine segment in a case
of obstructed delivery. B R
—Bandl's ring. (Modified
from Schroeder).

so rare as is usually believed. Any portion of the
uterus may rupture, but, with a very few exceptions,
the rupture always commences in the thinned lower
uterine segment (v. page 57) (v. Figs. 30 and
31). Starting from there, it may extend in any
direction—upwards towards the fundus, downwards
towards the vagina, or circularly round the uterus.
In this last case the entire lower uterine segment
may be torn off. A distinct variety of rupture, viz.
rupture by attrition, or rubbing through, of a portion
of the uterine wall, is sometimes met with. This
particularly happens in cases of contracted pelvis,
where the uterus may become caught between the
descending head and the promontory of the sacrum.
In these cases a circular hole may be completely
rubbed through the wall of the uterus ; or, more
commonly perhaps, the vitality of the compressed
portion may be so destroyed that it sloughs away
after delivery. There are two degrees of rupture :—

(1) Complete.

(2) Incomplete.

In the first the laceration extends through the
uterus and the investing peritoneum. In the
second the peritoneum is intact, and accordingly
there is no communication between the uterus and
the peritoneal cavity.

Ætiology.—The chief causes of ruptured uterus
are :— (1) Obstructed delivery from any cause, as,
—contracted pelvis, hydrocephalic
head, tumours blocking the pelvis, &c.

(2) Fatty degeneration of the uterus.

(3) A weak cicatrix resulting from a former
Cæsarean section.

(1) In obstructed delivery rupture always commences in the lower uterine segment. This is easily understood when we remember the result of the retraction of the muscle fibres by which the fundus becomes thicker, at the expense of the lower uterine segment (*v.* page 57).

(2) In fatty degeneration rupture may occur in any part of the uterus. It may occur at the commencement of labour, and cannot be foreseen.

(3) The cicatrix, resulting from former Cæsarean section, may rupture at the next pregnancy, if it be not firmly united.

Symptoms.—It is best to consider the symptoms of rupture of the uterus under three heads, viz. :—

(1) Threatened rupture.

(2) Sudden rupture.

(3) Gradual rupture.

(1) The symptoms of threatened rupture of the uterus are:—a rising temperature—above 101° F.; an increasing pulse rate—more than 110 per minute; continuous or tonic uterine contractions; Bandl's ring felt more than 1½ inches above the symphysis; ballooning of the vault of the vagina; standing out and tenseness of one or both round ligaments.

(2) The symptoms of sudden rupture are:—a sensation as if something had burst internally; cessation of labour pains; recession of the presenting part, unless it be already fixed; collapse, rapid pulse, falling temperature, all in proportion to the amount of hæmorrhage that is occurring; intense pain over the abdomen. These are the classical symptoms, but any or all of them may be absent.

(3) Gradual rupture. This is the commonest

manner in which rupture occurs, and its symptoms
are ill-defined. Nothing abnormal may be noticed
until the time comes to remove the placenta, when
upon introducing the hand into the uterus the rent
is discovered. If there be hæmorrhage, there will,
of course, be the symptoms of collapse. If rupture
be so extensive that the child escapes into the abdomen, the empty uterus may be felt by abdominal
palpation lying tightly contracted at the pelvic brim;
and the fœtal parts will be felt with unusual dis-
tinctness.

Treatment.—The treatment is prophylactic or
active as the case may require. The prophylactic
treatment consists in correcting malpresentations of
the child or obliquity of the uterus, and in immediate
delivery, if the indications of threatened rupture
appear. If the anterior lip descend in front of the
head, and become caught between it and the symphysis, it must be pushed above the convexity of
the head, and kept there during a pain. It will
then remain up of itself.

The active treatment depends entirely upon the
condition of affairs present. If the child be undelivered when the rupture is diagnosed, it must
be delivered at once. If it be in the uterus,
apply forceps or perforate; if it have escaped into
the abdomen, laparotomy is necessary. If there be
much hæmorrhage from the laceration, the uterus
must be removed. If the child be already delivered
before the rent is noticed, the treatment to be
adopted depends upon the amount of hæmorrhage.
Remove the placenta; and, if there be no hæmorrhage to signify, pass a strip of gauze through

the rent, so as to allow drainage.   No further treatment is necessary.   The gauze should be removed
in twenty-four hours.   If there be much hæmorrhage,
laparotomy followed by removal of the uterus is
indicated.   The usual operation in these cases is
supra-vaginal amputation of the uterus with extra-
peritoneal treatment of the stump.   But, in some
cases, it may not be possible to get the noose of the
serre-nœud sufficiently far down upon the uterus to
include the bleeding point.   In these cases total
extirpation of the uterus is the better operation.

Dangers.—The dangers of rupture of the uterus
are :—

  (1) Hæmorrhage.

  (2) Sepsis.

If the former occur it must be treated as described.
The latter will not occur in a healthy patient, if one
be a habitually aseptic accoucheur.

LACERATION OF THE CERVIX.

Lacerations of the cervix are seldom recognised
unless they cause hæmorrhage.   Their treatment in
that case is described under Post-partum Hæmorrhage.

LACERATION OF THE PERINÆUM.

This is one of the commonest accidents occurring
in midwifery.   It occurs far more frequently than
is supposed ; as, unless it be looked for with care,
it may not be noticed.   There are two degrees of
laceration of the perinæum :—

(1) *Complete*, where the laceration extends right through the perinæal body into the rectum.

(2) *Incomplete*, where the laceration involves the perinæal body alone.

Treatment.—Lacerations of the perinæum must be sutured immediately, for two important reasons :—

(1) To avoid the formation of a puerperal ulcer (*v.* page 271).

(2) To guard against future prolapse of the uterus. A deep laceration of the perinæum almost always involves the levator ani muscle. If this remain ununited the anterior vaginal wall has lost

FIG. 32.—Complete laceration of perinæum with continuous catgut suture in rectal and vaginal tear. (Modified from 'The Norris Text-book of Obstetrics.')

its support, and the integrity of the floor of the pelvis is destroyed.

The operation for a complete laceration is as follows :—The first step consists in turning the complete laceration into an incomplete laceration, by suturing the rent in the anterior rectal wall. This may be done, if the rent be small, by means of a purse-string suture which runs round the laceration. If the tear in the rectal wall be large, it is better to suture it by means of a continuous cat-gut suture. This afterwards is buried when the remainder of the perinæum is sutured (v. Fig. 32).

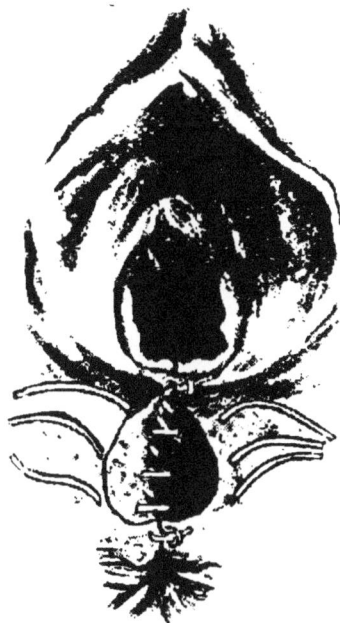

Fig. 33.—Complete laceration of perinæum turned into an incomplete laceration by suturing rectal tear. Perinæal sutures in position. (Modified from ' The Norris Text-book of Obstetrics.')

The complete laceration is now turned into an incomplete laceration, and it is in turn sutured by stitches passed from the external aspect of the perinæum. They are entered at the side of the laceration, passed completely beneath it, and brought out at the corresponding point upon the other side. They must be tied from behind forwards; and incomparably the best material to use is silkworm gut, as it does not absorb the discharge. If the laceration extend for any considerable distance up the posterior vaginal wall, it must be stitched separately with a continuous catgut suture, the stitches being passed from the vagina. The accompanying figure shows the method of procedure better than any words (v. Fig. 33). The stitches should be removed on the seventh day, unless they be catgut, which is absorbed.

CHAPTER XXVI.

CONTRACTED PELVIS.

Diameters of the Normal Pelvis—Varieties of Contracted Pelvis—
Ætiology—Diagnosis — Pelvimetry—Skutsch's Pelvimeter—
Symptoms—Mechanism of Pelvis Justo-minor—Mechanism
of Flat Pelvis—Mechanism of Generally Contracted Flat
Pelvis — Treatment — Table of Degrees of Contraction—
—Walcher's Position—Time at which to induce Premature
Labour—Müller's Method.

THE pelvis is said to be contracted if any of its
diameters be smaller than the normal. The normal
diameters of the brim of the pelvis measure :—

Conjugata vera . . 4—4¼ inches.
Oblique diameters (2) . . 5 ,,
Transverse. . . . 5¼ ,,

Varieties.—I intend to deal only with the common
forms of contracted pelvis. They are :—

(1) *Pelvis æquabiliter justo-minor.*—All diameters
are smaller than the normal, but preserve their
correct relation to one another.

(2) *Simple Flat Pelvis.*—The conjugate diameter
alone is narrower, the others are normal.

(3) *Generally Contracted and Flat Pelvis.*—All the
diameters are lessened, but the conjugate is narrowed
out of proportion to the other diameters.

For an exact description of the anatomical

peculiarities of these different varieties the reader is referred to Winckel's ' Text-book of Midwifery.'

Ætiology.—The *pelvis æquabiliter justo-minor* may be found in dwarfs, also in normally-sized women, in which case it is frequently associated with undeveloped sexual organs. The *simple flat pelvis* and the *generally contracted and flat pelvis* occur as the result of rickets, or may occur independently of that condition. In the latter case they have been attributed to carrying heavy weights during childhood.

Diagnosis.—The diagnosis of contracted pelvis is made,—from the appearance of the patient, from the history of the patient, from the symptoms of the patient during pregnancy and labour; and is confirmed by measuring the pelvis.

The appearance of the patient suggests contracted pelvis if any of the following conditions be present :—

> (1) Lateral curvature of the spine, particularly when it occurs in the lumbar region.
> (2) Marked lordosis.
> (3) Crooked legs, or legs of unequal length.
> (4) Enlargements of the junction of the cartilage and rib.
> (5) Pendulous abdomen.
> (6) Diminutive stature.

The history of the patient should be inquired into :—

(1) As regards her childhood; to ascertain if there be any evidence of early rickets, as,—late dentition, inability to walk at the proper age.

(2) As regards her previous labours; to ascertain whether they have been difficult or easy, whether the children were born dead or alive.

The symptoms of the patient during pregnancy and labour are of great importance. A contracted pelvis may commence to cause trouble in the early months of pregnancy; the fundus of a retroverted uterus may become incarcerated beneath the overhanging promontory (*v.* page 130). In the later months the growing uterus is pushed up out of the pelvis by the narrow brim; and, as a result of the lack of support which it thus experiences, it falls forward against the abdominal walls. A pendulous abdomen is thus produced. Also malpositions of the child are common, as I have explained before (*v.* page 42). When the patient comes into labour the head is found to be free above the brim at a time at which it ought to be fixed (*v.* page 47). Labour is very tedious, due to one or to all of the following reasons :—

(1) The narrow brim prevents the head from descending.

(2) A malpresentation may be present.

(3) The anteversion of the uterus prevents the due amount of help being obtained from the abdominal muscles.

(4) As a result of the head not filling the lower uterine segment the membranes rupture prematurely, and the liquor amnii drains away early in the first stage (*v.* page 89). The dilating action of the bag of membranes is thus lost, and the head itself is obliged to dilate the os. This it accomplishes slowly if it

come through the brim. If it do not come through, the os never completely dilates.

The uterine contractions are at first strong. If they continue so and do not succeed in delivering the child, the uterus will rupture. On the other hand, in many cases secondary uterine inertia sets in, owing to the obstructed delivery.

If the result of our examination suggest the possibility of contracted pelvis, then the pelvic diameters must be measured.

Pelvimetry may be external or internal. Very little information can be got from the former; the most valuable information from the latter. There are three points of some slight value which can be ascertained by external pelvimetry :—

(1) The length of the external conjugate, *i. e.* the distance between the upper edge of the symphysis externally, and the depression under the spinous process of the last lumbar vertebra. It is normally about 8 inches in length. If in any case it be found to be less than $6\frac{1}{4}$ inches, there is certainly some degree of contraction present.

(2) The normal distance between the anterior superior spines of the ilia is $10\frac{1}{4}$ inches; between the summits of the iliac crests $11\frac{1}{2}$ inches. Considerable shortening of these distances points towards contracted pelvis.

(3) The normal ratio of the distance between the spines, and the distance between the crests, is as $10\frac{1}{4}$ to $11\frac{1}{2}$. If the distance between the spines is either equal to or greater than the distance between the crests, the case is probably one of flat pelvis.

Internal pelvimetry is of much greater value than

external pelvimetry. By it we can measure the actual length of the *conjugata vera* and of the transverse diameter. This can be done by the fingers, or by means of Skutsch's pelvimeter. By the fingers we can only measure the oblique conjugate, and we have then to estimate the true conjugate from it. To measure the oblique conjugate, place the patient on her back, and under an anæsthetic if possible. Introduce the index and middle fingers into the vagina, and pass them upwards until the promontory of the sacrum is reached. While the fingers are in this position, mark, with the nail of the index finger of the other hand, the spot at which the subpubic

FIG. 34.—Diagram representing the manner in which the relationship between the true and the oblique conjugate is affected by the height of the promontory. (Modified from 'The Norris Textbook of Obstetrics.')

16

ligament crosses the index finger of the measuring
hand. Then withdraw the fingers, and measure the
distance between the tip of the middle finger and the
mark on the index finger. This is the oblique conjugate. To obtain the true conjugate, half an inch
must be subtracted. This is the average amount it
is necessary to subtract; the exact amount differs in
every individual case. If the symphysis lie more
horizontally, or if the promontory be lower than is
normal, half an inch will be too much to subtract.
If the symphysis lie more vertically, or if the promontory be higher than is normal, half an inch will
be too little (v. Figs. 34 and 35). Therefore, in
order to ascertain the exact amount that it is
necessary to subtract, we must allow for :—

FIG. 35.—Diagram representing the manner in which the relationship
of the true and the oblique conjugate is affected by the obliquity
of the symphysis. (Modified from 'The Norris Text-book of
Obstetrics.')

(1) The obliquity of the symphysis.

(2) The height of the promontory.

Skutsch's pelvimeter if carefully worked gives far more reliable information than the fingers (*v.* Figs. 36 and 37). It consists of three parts—a rigid limb with a slight curve upon it, a flexible limb, and a circular moveable bar which connects the two. The rigid and the flexible limbs lock into one another, in such a manner, that either the concave or the convex

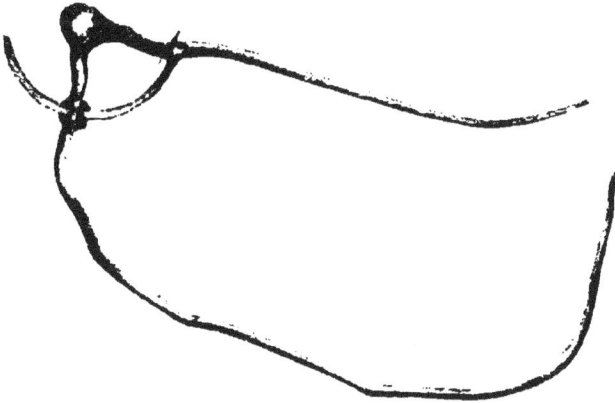

Fɪɢ. 36.—Skutsch's pelvimeter.

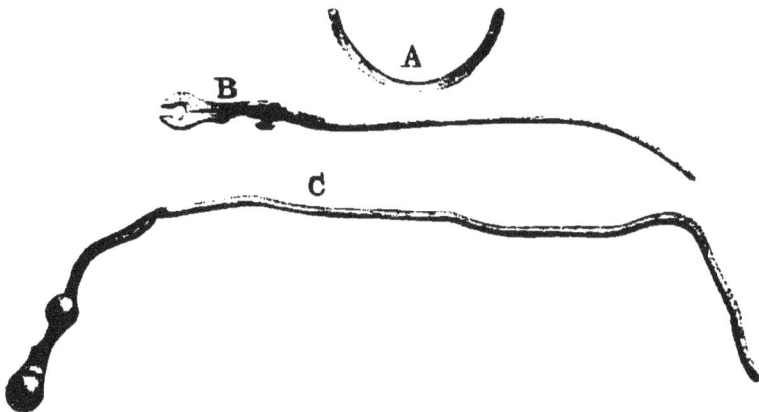

Fɪɢ. 37.—Skutsch's pelvimeter. A, moveable connecting bar; B, rigid limb; C, flexible limb.

aspect of the rigid bar can be turned towards the
flexible bar.   The moveable connecting bar is so
adjusted that the limbs can be separated from one
another, and then returned to exactly the same
position.   This is necessary in order to facilitate its
removal from the pelvis.

In order to use the pelvimeter, the patient
requires to be under an anæsthetic.   To measure
the *conjugata vera* a mark is made with an aniline
pencil on the skin over the centre of the symphysis
(*v.* Fig. 38).   The instrument is then so adjusted

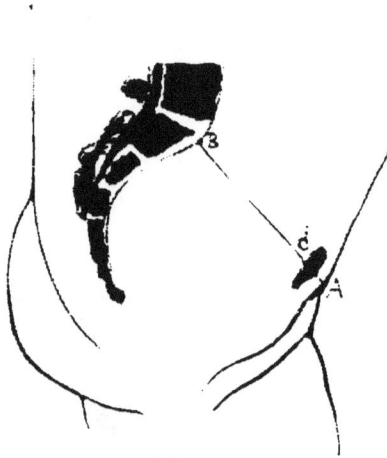

FIG. 38.—Diagram representing the distances measured, when ascertaining the length of the C.V.

that the rigid limb curves away from the flexible
limb.   Pass two fingers of the right hand into
the vagina, and upwards until they lie against the
promontory of the sacrum.   Then slip the rigid
limb of the instrument upwards, along the fingers,
until it rests on the most prominent point of the pro-

montory. Hold it exactly in this position while an
assistant bends the flexible limb, until it just touches
the blue mark over the symphysis. The instrument is
then carefully withdrawn, and the distance between
the extremities of the limbs measured (A B, Fig. 38).
Next reverse the rigid limb, so that it curves towards
the flexible one. Introduce the fingers again into
the vagina and feel for the most prominent point of
the back of the symphysis. Guide the rigid limb up
until it rests on this point; and hold it there while
the assistant bends the flexible limb, until it presses
against the blue mark with the same degree of force
as it did when taking the first measurement. Separate the limbs before removing the instrument, as
otherwise they might be forced apart. Then
remove it, and adjust the limbs to their original
position (A c, Fig. 38). Subtract the distance
between them, i. e. the thickness of the symphysis,
from the former measurement, and the result will
be the length of the true conjugate.

In order to measure the transverse diameter, make
a mark over the great trochanter of the femur at one
side (v. Fig. 39). With the rigid limb in the vagina,

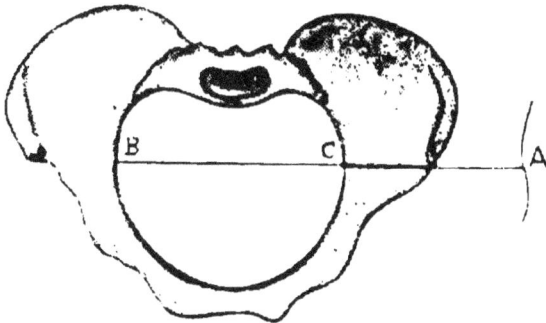

FIG. 39.—Diagram representing the distances measured, when ascertaining the length of the transverse diameter.

measure the distance from this mark to the most
distant point of the pelvic brim, at the opposite
side (A B, Fig. 39). Then, in the same manner,
measure the distance from the blue mark to the
nearest point of the pelvic brim, on the same side as
the mark (A C, Fig. 39). Subtract the measurement
thus obtained from the first measurement; the result
is the transverse diameter. From these measurements we can deduce the nature and degree of the
contraction.

Mechanism.—The mechanism of a vertex presentation differs considerably, in a case of contracted
pelvis, from the mechanism which occurs in the case
of a normal pelvis. The mechanism also varies in
the different forms of contracted pelvis.

Pelvis æquabiliter justo-minor.—The most important point in the mechanism of this variety of pelvis is,
that the normal flexion of the head is exaggerated;
consequently, the small fontanelle lies relatively
deeper in the pelvis than usual. With this exception, the mechanism is the same as in a normal pelvis.

Flat Pelvis.—The head engages with its sagittal
suture lying in the transverse diameter of the pelvis.
It then rotates round its occipito-frontal diameter,
in such a manner, that the sagittal suture approaches
the promontory of the sacrum, and the anterior parietal
bone presents. This lateral deviation of the head is
known as the obliquity of Naegelé; it is excessively
marked where there is much antero-posterior contraction of the pelvis. It is said, that, if the sagittal
suture come within three-quarters of an inch of the
promontory of the sacrum, the degree of contraction
of the pelvis must be so great that delivery is im-

possible. In addition to Naegelé's obliquity another rotation of the head occurs, this time about its bi-parietal diameter. One parietal bone rests upon the symphysis, the other upon the promontory; and the head rotates round these points, in such a manner, that, first the large fontanelle descends and passes through the brim, and then, rotation in the oppo-site direction occurring, the occiput descends. By this means the large diameters of the head pass the conjugate.

Generally Contracted and Flat Pelvis.—The mecha-nism in these cases is a combination of the mechanism that occurs in a justo-minor and in a flat pelvis. The head enters in the transverse diameter of the pelvis. Marked flexion occurs, so that the small fontanelle is situated more deeply than usual. Naegelé's obliquity also occurs, so that the anterior parietal bone comes first through the brim.

Treatment.—Four degrees of contracted pelvis are met with. Together with their appropriate treatment, they are as follow :—

Degree.	Measurement of C.V.	Treatment.
1st	3½—4 inches	Prophylactic version; or leave to nature,—*i. e.* allow the head to mould through the brim.
2nd	3—3½ inches	Premature labour; sym-physiotomy; perfora-tion; or Cæsarean section.
3rd	2—3 inches	Symphysiotomy; per-foration; Cæsarean section.
4th	below 2 inches (absolute contraction)	Cæsarean section.

(1) In the first degree of contracted pelvis we have a choice between prophylactic version, and leaving the head to mould. I do not include the application of forceps as a mode of treatment; because, if forceps will bring a head through a contracted brim, the contractions of the uterus will also bring it through, with less danger to mother and child. When we make up our mind to allow the head to mould through the brim of itself, we leave the case absolutely to nature, until either the child dies, or until danger to the mother appears. If the child die, there is no object in waiting any longer ; perforate the head and extract. If symptoms of danger to the mother show themselves, we may give forceps a trial, on the supposition that the contraction may not be so great as we think. If it fail, we must perforate. Prophylactic version consists in turning the child by external version into a breech presentation, at the commencement of labour. This is done in pursuance of the fact that the head moulds better when compressed from below upwards, i. e. as an after-coming head, than it does when compressed from above downwards, i. e. as a forecoming head. But, on the other hand, it must be remembered, that, when the head comes first, it may take an indefinite number of hours to come through the brim without detriment to the child. When the head comes last, it must be dragged through the brim in at most one minute, or the child will die of asphyxia. It is probably better to adopt moulding in this degree of contraction. In support of this I give Winckel's statistics :—

Method of Delivery.	Fœtal Mortality.	Maternal Mortality.
Moulding . .	12 per cent. . .	*
Forceps . .	27·7 ,, . .	4 per cent.
Version . .	40·8 ,, . .	3 ,,

(2) For the second degree of pelvic contraction premature labour is undoubtedly the best treatment. If we do not see the patient sufficiently early in pregnancy to adopt this mode of treatment, then symphysiotomy is to be preferred. The great objection to the latter is the number of assistants it requires, and the difficulty of the after-treatment. Cæsarean section on this account is usually to be preferred in private practice. If neither of these can be performed the child must be perforated.

(3) For the third degree of pelvic contraction symphysiotomy is the best treatment for its upper limits, i. e. for a *conjugata vera* of 2¾ inches or more; below that Cæsarean section, as far as we know at present, is safer. Perforation must be adopted if neither of these can be employed.

(4) For the fourth degree of pelvic contraction, i. e. absolute pelvic contraction, Cæsarean section is the only possible mode of delivery. Extraction of even a mutilated child is too dangerous an operation to be undertaken.

Walcher's position may be of considerable use, in any case in which a slight enlargement of the conjugate diameter of the brim is required. It consists in placing the patient in the dorsal position; with her

---

* If the treatment of moulding be properly carried out there should be no maternal mortality; as, when danger to the mother appears, the child is at once extracted.

hips so over the edge of the bed that her legs hang freely down, without any support. The lower portion of her body then rests upon the sacrum, and the weight of the unsupported lower limbs is transmitted through the ilio-femoral (Y-shaped) ligament to the pelvis. The movement which the sacro-iliac joints allow, permits as much of the pelvis as is formed by

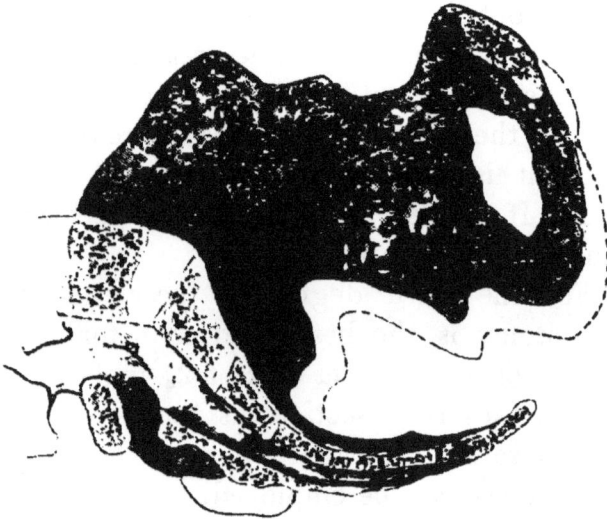

FIG. 40.—Diagram showing the increase in the C.V., brought about by Walcher's position. The dotted outline represents the new position of the pelvis, when dragged downwards by the weight of the limbs. (Slightly modified from Fothergill.)

the innominate bones to be drawn downwards by the weight of the limbs, as if it were rotating round the sacro-iliac joints. In this way the symphysis comes to lie at a lower level than is usual, so increasing the *conjugata vera* (*v.* Fig. 40). The average increase in the latter is about two fifths of an inch (Fothergill).

The correct time at which to induce premature labour, for the different degrees of contractions, is shown in the following table :—

Conjugata Vera.	Time to Induce Labour.
3 inches	30th week.
3¼ ,,	32nd ,,
3½ ,,	36th ,,

This manner of ascertaining the date at which to induce labour is open to two objections. In the first

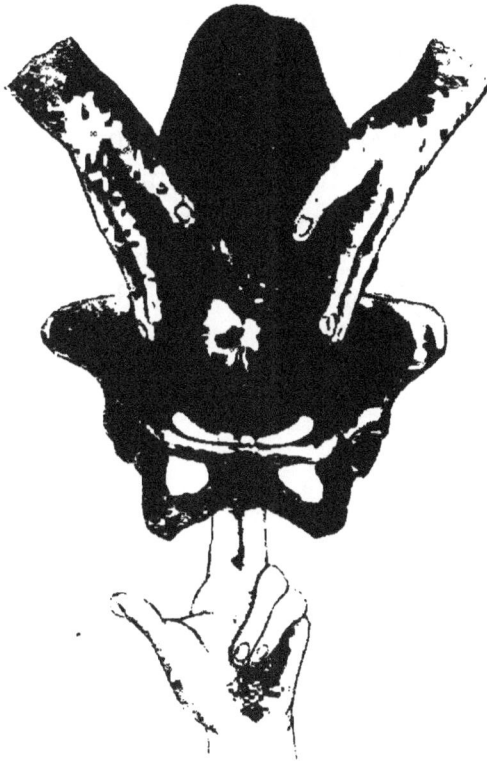

FIG. 41.—Müller's method of ascertaining the date at which to induce labour.

place, it is extremely difficult to be certain that we are correct in our calculations of the duration of pregnancy. In the next place, even if we can tell the exact age of pregnancy, this table makes no allowance for the varying size of the child's head. *Müller's method* of ascertaining the date at which to induce labour is much more exact; and allows both for the degree of contraction of the pelvis, and for the size of the child's head. It is carried out as follows :— Place the patient in the cross-bed position, or upon a Schroeder's gynæcological chair. Introduce two fingers into the vagina and palpate the presenting head. Then get an assistant to grasp the head through the abdominal wall, and to endeavour to push it down through the pelvic brim (*v.* Fig. 41). If he succeed, it is too soon to induce labour. This manipulation should be performed, at intervals of a few days, until the day comes that he cannot push it through the brim. The first day on which this occurs, is the day on which labour should be induced.

CHAPTER XXVII.

PROLAPSE OF THE CORD.

Difference between Presentation and Prolapse of the Cord—
Ætiology — Diagnosis — Treatment: Postural, Reposition,
Version, Forceps.

By the term presentation of the cord, is meant that
the cord lies in front of the presenting part, the
membranes being unruptured. Prolapse of the cord
is the term applied to the same condition after the
membranes have ruptured.

Ætiology.—The commonest cause of presentation
or prolapse of the cord, may be stated generally to be
any condition that interferes with the normal adap-
tation that exists between the presenting part and
the lower uterine segment. The chief of these conditions are :—

(1) Contracted pelvis.
(2) Malpresentations :—face, breech, cross-
birth, brow.
(3) Hydramnios.
(4) Twins.

In any of these conditions the presenting part may
not fill the lower uterine segment. Consequently
when the membranes rupture, the liquor amnii comes
away with a rush, and may carry down a loop of the
cord with it. Prolapse may also occur owing to :—

(5) Low attachment of the placenta, *i. e.*
placenta prævia.

(6) An abnormally long cord.

(7) Marginal insertion of the cord into the
placenta, *i. e.* battledore placenta.

Diagnosis.—The diagnosis is obvious in prolapse
of the cord. A loop of it can be felt in the vagina,
or may even be seen protruding from the vulva. In
presentation of the cord, its coils are felt in front of
the presenting part ; and, if the child be alive, the
cord pulsates.

Treatment.—The cord must be replaced, or the
child rapidly delivered if its life is to be saved. If
the cord be not pulsating when the condition is discovered, and if the fœtal heart cannot be heard, the
child is dead, and there is no need to interfere.
It must always be remembered that the cord stops
pulsating at least a minute before the death of the
child ; consequently, if we know that the pulsations
have only just ceased, we should deliver at once, if
possible, and not give the case up as hopeless.

The treatment of the case varies with the conditions
present. If it be a case of presentation of the cord,
and the presenting part be not fixed, try the *postural
treatment.* Place the patient in Trendelenburg's position, *i. e.* with the buttocks raised and the head low ;
the fœtus will then tend to fall towards the fundus
under the influence of gravity, and the cord may do
the same. Examine vaginally while the patient is
still in the same position. If the cord have gone up,
push the presenting part into the brim of the pelvis,
rupture the membranes, and then allow the patient
to lie down again. Keep the presenting part pressed

into the brim until a contraction fixes it, or apply a
tight abdominal binder with the same object.   An ex-
tempore Trendelenburg's table can be made by placing
an ordinary square kitchen chair on its face along the
bed, covering it with pillows in order to protect the
patient.   By this means the body is placed on an
inclined plane, which is just as efficacious and far
more comfortable for the patient than the knee-chest
position.

If this method do not succeed, or if the membranes
be ruptured, an attempt must be made to replace
the cord.   This is an extremely difficult operation to
perform ; as fast as we replace one loop of the cord
another comes down.   The patient should be placed
under an anæsthetic, as any straining renders the
operation impossible. Then grasp the cord in the hand,
carry it up past the presenting part, and endeavour
to hang it over a limb.   As the hand is brought down,
press the presenting part down into the brim from
without.

If the os be not sufficiently dilated to enable the
hand to be introduced, or if we fail to replace the
cord with the hand, a repositor of some kind may be
used.   The best form of repositor is made as follows :—
Take a No. 10 or 12 gum-elastic catheter with a
stilette.   Knot together the ends of a piece of sterilised silk about seven inches in length.   Pass any part
of the loop thus formed through the eye of the catheter
and push the stilette up in such a way that it passes
through the loop.   The instrument is then ready.
To use it, pass the loop of silk, that hangs from
the eye of the catheter, round the prolapsed portion
of the cord, and then throw the loop over the top of

the catheter (*v.* Fig. 42). Pass the catheter upwards into the uterus until the cord is above the
presenting part, and then withdraw the catheter
gradually, at the same time pressing down the

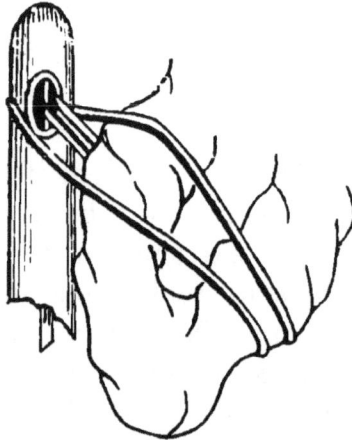

FIG. 42.—Catheter used as a repositor for the cord, showing the
manner in which the string is adjusted.

presenting part into the brim. So long as the
catheter is pushed up, the loop cannot slip off
the top of it ; but, as soon as we commence to
withdraw it the loop slips off the top, and the cord
is set free.

If reposition fail, and the head be presenting,
turn the child into a breech presentation, and
draw down a foot. By this manœuvre, we obtain a
presentation which is not so likely to press upon the
cord as is a head presentation ; inasmuch as the
breech does not so completely fill the lower uterine
segment. The case must be watched very carefully ; and, if the cord stop pulsating, we must extract the child at once. Version is performed by

the internal or the bipolar method, according to the size of the os.   There is one point in which the method of performing version, in a case of prolapsed cord, differs from the usual method.   The child should be turned by pushing the head in the direction of its abdomen, *i. e.* in the reverse of the ordinary direction.   The object of this proceeding is to keep the umbilicus of the child as far away from the os uteri as possible, in order to prevent more cord from prolapsing.

If version cannot be performed, *i. e.* if the head be fixed, the child must be extracted immediately by forceps.   If the os be sufficiently dilated, there will be no difficulty in the proceeding.   If it be not dilated, it must be incised bilaterally, and forceps applied.   Never drag the head through an os which is not sufficiently dilated to allow it .to pass through without tearing.   A laceration will inevitably occur, and a laceration once started may extend into the lateral fornix and open the uterine artery.   If incisions be made, they must be stitched immediately after delivery.

While the necessary preparations are being made for any operation for prolapse of the cord, the patient should be placed on the side at which the cord lies in a head presentation, or upon the opposite side in a breech presentation.   The reason for this proceeding has been explained (*v.* page 113).   Also the patient should be told not to strain or " bear down."

17

CHAPTER XXVIII.

ECLAMPSIA.

Definitions — Morbid Anatomy — Ætiology : Frerich's Theory,
Traube-Munk-Rosenstein Theory, Stumph's Theory, Toxine
Poisoning—Predisposing Conditions—Symptoms : Prodromal,
Actual—Complications—Treatment : Prophylactic, Active,
Chloroform Treatment, Morphia Treatment, Position of
Patient, Feeding, Induction of Labour, Venesection—Pro-
gnosis.

THE term eclampsia is applied to a form of convulsive attacks which are peculiar to women during
pregnancy and the puerperium.

Morbid Anatomy.—If a post-mortem examination
be made on a woman who has died of eclampsia, a
series of more or less constant morbid conditions is met
with. Nothing, however, has been found which can
be regarded in the light of a primary lesion. The
kidneys are diseased in almost all cases ; most frequently there is an acute congestive catarrh, more
rarely a diffuse interstitial nephritis. The *brain* is
anæmic, with a certain amount of œdema and consequent flattening of the cerebral convolutions
(Winckel). The bases of the *lungs* are markedly
œdematous. The *liver* sometimes exhibits signs of
fatty degeneration.

Ætiology.—We have very little positive knowledge of the causation of eclampsia. Four theories require special mention :—

(1) *Frerich.*—That eclampsia is uræmic in origin, *i. e.* is due to the retention of urea in the blood. In opposition to this theory is the fact, that in the case of patients who die from eclampsia, no storage of urea can be found in the liver or muscles. Again, in the case of those who recover, no increased amount of urea is excreted in the urine.

(2) *Tranbe-Munk-Rosenstein.*—That eclampsia is primarily caused by the hydræmia of the blood which occurs during pregnancy ; the secondary cause being furnished by the contractions of the uterus, which, by raising the blood-pressure, cause œdema of the brain, it in turn causing anæmia. In opposition to this is the fact, that hydræmia is not particularly well marked in eclamptic patients. Also fits occur before the onset of labour, *i. e.* before the blood-pressure is raised (Winckel). In opposition to this last statement may be urged the fact, that perceptible painless contractions of the uterus occur in the later months of pregnancy. If these contractions be strong enough to cause a very perceptible hardening of the uterus, they must also be sufficiently strong to raise the blood-pressure.

(3) *Stumph.*—That the fits are due to the circulation, in the blood, of some poison produced by an abnormal decomposition in either mother or child. This poison Stumph says may be acetone,—a substance which he has almost always detected in the urine of eclamptic patients. That this poison, in its passage through the kidneys, causes nephritis ; through the

liver, a destruction of the parenchyma of that organ; through the brain, convulsions and coma.

(4) The theory most generally accepted at present is, that the fits are due to the retention of some poison or poisons, which under normal circumstances are excreted in the urine. Coincidently with the onset of eclampsia the urine has been noticed to contain a diminished quantity of these substances; the total amount of urine passed is also considerably diminished. Coincidently with the recovery of the patient, the quantity of toxic substances in the urine is considerably increased, as is the total amount of urine passed. It is uncertain what the exact nature of these toxines is.

Apart from theories, there are certain conditions which are known to predispose to eclampsia. These are :—

(1) Acute and chronic diseases of the kidneys, particularly that form known as " pregnancy kidney."

(2) Obstructed delivery.

(3) Old and very young primiparæ,—*i. e.* rigid uterine muscle fibres.

(4) Long retention of the excretions.

(5) Multiple pregnancy.

Symptoms.—The symptoms must be considered under two heads :—(1) prodromal; (2) actual.

The prodromal symptoms come on a short time before the onset of the fits, and are of great importance. The timely recognition and treatment of them may stave off the threatened attack. They are :—complete or partial, temporary or persistent, loss of vision; flashes of light before the eyes; vertigo; drowsiness; mental depression; nausea; and epigastric pain. At the same time the amount of urine excreted becomes very considerably diminished; and, if a specimen can be obtained for examination, it is found to contain a very large quantity of albumen, and numerous granular and fatty tube casts.

The actual symptoms commence with the onset of the fits, which may occur quite suddenly. A fit lasts from a half to one minute, and consists of three stages, a preliminary stage, a tonic stage, and a clonic stage,—followed by a varying period of coma. In the preliminary stage, the eyelids twitch vigorously and spasms of the muscles of respiration occur. Then the tonic stage supervenes, and the patient lies with all her muscles contracted. She becomes deeply cyanosed, and froth appears at the mouth. The clonic stage follows; she " works " vigorously for a time, then respiration gradually returns, and the patient falls into a condition of deep coma. The duration of coma varies according to the number of fits that have occurred. At first it may only last a few minutes; but, as the number of fits increases, she lies in a continuous condition of coma during the intervals between them. The number of fits varies from one or two up to any number. Winckel says he has seen a patient recover after having had eighteen. They may pass off entirely for a time, and then recur. In a severe case the fits follow one another at ever-shortening intervals; the heart becomes weaker; and the lungs œdematous, at first at the bases, and then universally. The pulse is frequent; the temperature, which was normal at first, rises as the case progresses, perhaps attaining a height of 104° F.;

total or partial loss of vision may persist for a
considerable period after the fits have ceased.

Complications.—The principal complications to be
feared are failure of the heart, and œdema of the
lungs. They occur in almost all fatal cases, and are
the direct cause of death. Hæmorrhage into the
brain may occur during a fit, or may happen even
after the fits have entirely ceased. In a case
which I attended the patient died of cerebral hæmorrhage, which apparently took place thirty-six hours
after the last fit.

Treatment.—The treatment of eclampsia must be
considered under two heads, prophylactic and active.

Prophylactic treatment should be adopted in the
case of every patient who has persistent albuminuria,
especially if there be tube-casts in the urine. The
patient should be placed on milk diet, and limited to
it as far as possible. Her bowels should be kept
free by the daily administration of a purgative such
as the Pil. Colocynth. et Hyoscyami (B. P.). The
amount of urine she passes must be most carefully
watched, in order that any marked diminution may
be immediately detected. If this occur, a hydragogue
purgative must be at once administered ; followed, if
the diminution in the urine be considerable, by a
wet pack or hot baths. After this the patient is
wrapped in blankets in order to favour sweating. A
suitable purgative to administer in these cases consists of Calomel 10 grains, combined with Pulv.
Jalapæ Co. 1 drachm, and followed in eight hours
by an enema if necessary. If in spite of all precautions an eclamptic fit occur, our treatment must then
become active.

The active treatment of eclampsia must be chiefly directed towards two points :—

　(1) The arrest of the fits.

　(2) The staving off of complications.

(1) The fits must be checked at the earliest possible moment, as each successive fit leaves the patient more comatose, and more likely to fall a victim to the complications of a failing heart and œdema of the lungs. Two different lines of treatment are recommended :—

(*a*) The first of these is the chloroform and chloral treatment. This consists in administering, upon the onset of the attack, thirty grains of chloral hydrate by the rectum ; while at the same time inhalations of chloroform are given whilst the fit lasts. The administration of the chloral is repeated every two hours until the fits cease, but not more than three and a half drachms should be given in the twenty-four hours. The administration of chloroform is commenced as soon as any sign of the onset of a fit occurs, and continued until the fit ceases.

(*b*) The second line of treatment consists in the administration of large doses of morphia, hypodermically, as recommended by G. Veit. It is considerably the better method of treatment, and is carried out as follows :—Half a grain of morphia is administered hypodermically upon the onset of the first fit, and is followed every two hours by the administration of a quarter of a grain, until the fits cease. Not more than two grains should be given in the twenty-four hours.

Either of these lines of treatment will check the fits ; but both chloroform and chloral depress the heart seriously, and consequently favour that most dreaded complication—heart failure. On this account the morphia treatment is to be preferred. In addition to either of these treatments the patient must be freely purged. If the patient be conscious, calomel and compound jalap powder, as recommended above, are the best purgatives ; if, however, she be comatose, it is useless to place any bulky medicine in her mouth, as it would not be swallowed. Two minims of croton oil, made into a small bolus with a little butter, and placed as far back upon the tongue as possible, may reach the stomach. A soap and water enema should also be given. Every effort should be made to encourage free sweating. The patient must be kept in blankets, and hot baths administered if possible ; if the latter be not possible, a wet pack may be tried instead. The amount of urine excreted must also be increased as far as possible ; hot stupes over the kidneys, and abundance of fluid by the mouth if the patient be conscious, will sometimes be of avail.

(2) Complications will be staved off most of all by intelligent nursing, and by paying the greatest attention to details. Whilst the patient is in the fit, she must be prevented from injuring herself. She is especially likely to bite her tongue, when it is extruded during the fit. This is prevented by the use of a gag. A very serviceable one may be made in a moment by wrapping a towel or other piece of cloth round a spoon. All feeding by the mouth must be stopped, as, if the patient be comatose, any food is more likely to find its way into the lungs than into the stomach. If it be required to give nourishment while she is in this condition,

nutrient enemata must be administered. The position of the patient must be such that the saliva, which tends to collect in the mouth, will trickle out at the side of it, instead of running down into the trachea, *i. e.* she must lie upon her side, and not upon her back. If the heart become weak and rapid, digitalin and strychnine may be administered hypodermically.

There are two or three points in the treatment of eclampsia which are still matters of dispute. The first of these is the question as to the advisability of the induction of premature labour and of immediate delivery. It is of course a fact, that, if the patient were not pregnant, the eclamptic fits would not occur. Hence the supporters of immediate delivery reason, that, if pregnancy be brought to an end, the fits will cease. There are two objections to this line of reasoning. First, that the effects of pregnancy on the maternal organism do not disappear the moment the child has passed the vulva. Secondly, that uterine contractions directly excite the convulsive attacks, in the same way that any other violent movement or emotion will excite them. If the fits can be checked before labour comes on, the prognosis of the case will be improved. If labour come on before the fits are checked, the shorter the duration of labour the better the prognosis of the case. Following this reasoning, I recommend not to induce labour; but, if it occur spontaneously, to shorten its duration as much as possible, without employing such undue violence as would cancel the good effects obtained by the lessened period of uterine contraction. In other words,—apply the forceps and deliver the child as soon as the necessary conditions for its application are present ; do not adopt such violent measures as Cæsarean section or *accouchement forcé*. Any operation, which is to be performed, must be performed while the patient is under an anæsthetic, as by so doing the shock of the operation is lessened.

The advisability of venesection is another unsettled point. The toxicity of the blood in eclampsia has been proved by Professor Chambrelent ; and it is thought by some that if the patient be freely bled, a certain amount of the toxic agent will be removed. On the other hand, free venesection may dangerously lessen the patient's strength. It may be a useful line of treatment in the case of a strong plethoric patient.

Prognosis.—The prognosis for the life of the infant in eclampsia is very grave. For the mother the prognosis varies according to the time at which the fits commence. The earlier in pregnancy or the later in labour that the fits commence, the bet er is the prognosis. The prognosis is worst when the onset of the fits occurs at the commencement of the first stage ; it is best when they start during the puerperium. The greater the number of fits, the worse is the prognosis. If the child die, the prognosis is improved. The amount of urine passed, and the quantity of albumen in it, are also important guides.

CHAPTER XXIX.

SEPTIC INFECTION.

Varieties of Septic Infection—Sapræmia: Ætiology, Symptoms, Prognosis, Treatment—Puerperal Ulcer: Treatment—Acute Sepsis: Ætiology, Prognosis, Treatment, Anti-streptococcic Serum—Pyæmia: Ætiology, Symptoms, Prognosis, Treatment.

SEPTIC infection manifests itself during the puerperium in three distinct forms :—

    I. Sapræmia or septic intoxication.

    II. Acute sepsis.

    III. Pyæmia.

1. SAPRÆMIA.

Sapræmia or *septic intoxication* is the condition which arises from the absorption of the products of decomposition. Saprophytic organisms are carried into the vagina or uterus by the air which gains admission during the third stage of labour. They lodge in any dead matter, as blood-clots or portions of placenta, and there generate ptomaines. These latter are absorbed by the patient, and sapræmia results.

Ætiology.—The direct cause of sapræmia is the entrance of air laden with saprophytes into the vagina or uterus. This occurs to some extent in almost every confinement; but sapræmia does not result, unless there be dead matter left in the uterus for the saprophytes to feed upon. Improper management of the third stage of labour, *i. e.* premature expulsion of the placenta, is the common cause of the presence of dead matter in the uterus. The lateral position in the third stage by favouring the entrance of air, and the insufficient control of the fundus by allowing the uterus to fill with clots, are also well-recognised causes of sapræmia.

Symptoms.—The symptoms set in from the third to the fifth day after the birth of the child, and usually commence gradually. The temperature rises to 101° F. or 102° F., and the pulse becomes proportionately rapid. At the same time the lochial discharge has a foul smell. If the case be treated, the symptoms disappear; otherwise, the temperature rises higher on the following night, and the patient may have a slight rigor. If the case be still untreated, the symptoms become very much more marked, and the patient feels very ill indeed. Bacteria, which at the commencement were saprophytic, and so were only able to live upon dead matter, will, under suitable conditions, become pyogenic and capable of existing upon living tissues. These suitable conditions occur in the case of a neglected sapræmia. Consequently, the bacteria which have gained in virulence by feeding on portions of placenta and clots, now attack the uterus itself and cause a septic endometritis. The chief symptom of this condition is a very foul-smelling and profuse discharge coming from a subinvoluted uterus.

The inflammation may then extend in two ways. It may infect a laceration about the cervix, and so travel outwards into the tissues of the broad ligament, thus causing a parametritic inflammation. It may extend into the tubes and so reach the peritoneal cavity. When the inflammation has reached the pelvic peritoneum, two terminations are possible. First, and more rarely, a general peritonitis may be set up, which will infallibly kill the patient. Secondly, a protective local peritonitis seals over the fimbriated extremities of the tubes, and thus prevents the inflammation extending beyond the pelvis. This is the commonest termination, and is the cause of subsequent adherent retroversions, fixed ovaries, and pyosalpinx.

From any of these conditions, viz. septic endometritis, salpingitis, or peritonitis, a general pyæmia may result. The extension of the inflammation from the uterus is shown by intense pain, accompanied by considerable elevation of temperature, and by the occurrence of rigors. The presence of inflammatory exudations may be determined by vaginal examination, or by palpation of the lower part of the abdomen.

Prognosis.—If sapræmia be treated in time, the patient practically always recovers. If the condition be untreated, she may die from ptomaine poisoning, or from pyæmia. If extension of the inflammation to the tubes take place, she may die of septic peritonitis ; and even in the most favourable cases she will be an invalid for a very long time, and perhaps for life.

Treatment.—The prophylactic treatment of sapræmia consists in the proper management of the third stage of labour. If sapræmia occur, it must be treated at once. When the symptoms first appear, raise the head of the bed slightly, and so favour free drainage from the vagina. With the same object administer a purgative, which, by causing bearing-down efforts, assists in emptying the vagina. If the third day be passed, the patient may be allowed to kneel up in bed when passing water. If in spite of this treatment the temperature still keep high, she must be given a copious vaginal douche of hot creolin solution. If the decomposition be limited to the vagina, the temperature will then fall. If it still remain high the uterus is probably infected ; accordingly a vaginal douche should be again administered, and followed by an intra-uterine douche. The former may be given with a glass nozzle, but for the latter a large-sized Bozeman's catheter must be used in order to permit of free return of the fluid. In the great majority of cases the temperature will now fall. If it still remain high, the uterus must be douched twice daily.

If the discharge remain foul in spite of two or three intra-uterine douches, the uterus should be carefully curetted with a blunt Rheinstadter's douche curette, in order to remove any portion of placenta or of membrane which has been left behind. The curetting should be thorough, but the greatest care must be taken that the uterus be not perforated. If it be obvious that the inflammation has extended to the tubes, curetting is probably contra-indicated. There is always the risk that it may set up a general pyæmia. If the tubes be infected, the worst that can occur has occurred, and consequently there is no compensation to make up for the attendant risk. If the inflammation extend beyond the uterus, it is treated by rest in bed, and the pain is relieved by hot abdominal stupes.

In all cases the diet of the patient must be attended to. She must get plenty of nutritious and digestible food. While the temperature is high, alcohol must be given in doses of from four to ten ounces in the day, in accordance with the severity of the case.

Puerperal ulcer is a condition which may accompany sapræmic infection, or may exist alone. It is an ulcer of varying size, with a grey sloughing base and an inflamed margin. It forms on lacerations of the genital tract; and it causes a profuse foul discharge, accompanied by a rise of temperature and slight constitutional disturbances.

Treatment.—The treatment is purely local, and consists in the careful washing away of all discharge and in the application of iodoform powder to the ulcer. Vaginal douching is contra-indicated, lest the putrid discharge be carried up into the uterus; instead of it, the head of the bed may be raised to favour free drainage.

It is often of great assistance to be able to distinguish healthy from putrid lochia, at a glance. This is done by examining the stain which the lochia leaves upon the sheet or diaper. Healthy lochia causes a stain resembling that left by a drop of blood. It is deep red in the centre, and fades away gradually into a purely serous margin. Putrid lochia causes an exactly opposite variety of stain. It has a hard, deeply stained edge, and the colour fades slightly towards the centre of the stain.

II. Acute Sepsis.

Acute sepsis is the term applied to the condition that results from the entrance of pyogenic bacteria into the circulation of the patient, *via* the uterine

lymphatics.   It is the most fatal disease to which
puerperal women are liable.

Ætiology.—Acute sepsis is directly due to the infection of the uterus with *Streptococcus pyogenes,*
which has been introduced by the fingers or instruments of the medical attendant or nurse.

Symptoms.—The symptoms appear from twenty-
four to fifty hours after inoculation.   They are
usually ushered in by a severe rigor, during which
the temperature rises to 104° F. or 106° F.   The
pulse is exceedingly frequent, and is even out of proportion to the temperature.   The rigors recur at short
intervals, and the patient is bathed in a profuse cold
sweat between them.   The secretions peculiar to the
puerperium, *i. e.* the lochia and milk, cease completely,
if they be ever established.   The patient looks extremely ill, and is absolutely sleepless.   Her face is
pinched, and has a sub-icteric tinge ; the angles of the
mouth and nose are drawn down ; and the eyes appear
sunken into the head.   A very common symptom is
extreme depression.   In some of the worst cases the
patient may say that she feels extremely well, and may
even wish to be allowed up (*v.* page 123).   This
condition is known as *euphoria,* and is due to the
fact that the higher centres are dulled by the
poison which is circulating in the system.   It is a
sign of the worst possible import.   A frequent
concomitant of the general infection of the patient,
is a diffuse septic peritonitis.   The duration of
the disease is at the most a week, frequently
only a couple of days.   The temperature rises
during the entire time, and may reach 106° F. or
107° F.   Towards the last the heart fails rapidly.

Prognosis.—The prognosis is absolutely bad. If the patient recover, it is most probable that the diagnosis was incorrect.

Treatment.—When the symptoms appear, the vagina and uterus should be douched, on the chance that they may be due to a local infection. If, however, the symptoms do not improve rapidly, it is useless to continue the douches. The only drug that seems to be of any avail is alcohol, and it should be pushed to the utmost extent. The patient must be given literally as much as she can be urged to take, and this is rarely more than sixteen to twenty ounces in the twenty-four hours.

During the last year, there has been a considerable amount of attention paid to the possibilities of anti-streptococcic serum. Many favourable instances of its use have been recorded, but it is as yet too soon to make any certain pronouncement as to its efficacy. Undoubtedly many cases, which have been diagnosed as acute sepsis, have recovered under its use ; but also many cases of purely sapræmic infection have been treated with it, under the belief that they were, or for fear that they might be, acute sepsis. And of course from the fact that such cases recovered, we can deduce nothing as regards the efficacy of the serum in the graver infection.

Professor Denys, of Louvain University, and Dr. Leclef have together made a careful investigation into its efficacy. From experiments made upon rabbits, and from the use of the serum in cases of streptococcic infection, they have come to certain conclusions. Briefly summarised, these are as follows :—

(1) That the majority of cases of puerperal *streptococcic* infection are benefited by the use of the serum. That a certain number of cases are not benefited, either because the serum is not sufficiently active, sufficiently recently made, or because the infection is due to some organism other than *Streptococcus pyogenes.*

(2) That to be of avail, a sufficiently large dose of the serum must be administered. Dr Denys recommends that from 50 c.c. to 200 c.c. be used in proportion to the severity of the case.

(3) That this dose must be administered at one time, or any rate within twenty-four hours.

(4) That there must be no hesitation in administering the serum at once, otherwise it may be too late.

(5) That the serum must be properly prepared ; that it must be as recently prepared as possible ; and that the infection must be due to *Streptococcus pyogenes.*

If the serum be used, strict precaution must be taken to ensure the asepsis of the syringe with which it is injected, and of the skin through which the puncture is made.

III. Pyæmia.

The chief difference between acute sepsis and pyæmia appears to be, that, in the former, the virulence of the invading bacteria is so great, that the patient dies before any gross pathological changes have time to occur in the body ; whilst in the latter, the virulence is so diminished, that although sufficient to set up symptoms of septic poisoning, still it is not sufficient to kill the patient before certain pathological changes have occurred. These changes

if carefully examined into will be most frequently
found to be directed towards the limitation of the
distribution of the poison, and so to be protective.

Ætiology.—Pyæmia is due to the infection of the
patient through the blood-vessels with pyogenic
bacteria, which have been introduced into the uterus.
The infecting bacterium, in most cases, is morpho-
logically the same as that which causes acute sepsis,
i. e. Streplococcus pyogenes ; doubtless mixed forms
of infection occur also. In most cases the bacteria
are at first lodged in the clots which fill the uterine
sinuses, and gain entrance to the general circulation
as the clots break down. Pyæmia may also result
from a neglected sapræmia.*

Symptoms.—The onset of the symptoms does not,
as a rule, take place until the tenth day after
delivery. The patient may have had an apparently
normal puerperium up to that date, or may have
suffered from sapræmic infection of the uterus. The
onset of the symptoms is marked by the occurrence of
a severe rigor, followed by a rapid elevation of temperature, up to 104° F. or 106° F. The pulse rate
increases proportionately. The rigor lasts a short
time ; in a few hours the temperature falls to normal,
and the patient may appear to be as well as she was
previous to the attack. Another rigor, however,
follows in from twelve to twenty-four hours, and is
followed by others at shorter intervals. Finally, the

---

* It must be remembered, that the entrance of staphylococci
and streptococci into the uterus does not necessarily result in
acute sepsis or pyæmia. In some cases the infection remains
local, causing a septic endometritis with its train of resultant
conditions, as described under sapræmia.

temperature remains continuously between 100° F.
and 106° F.

In from three days to a week after the onset of
the symptoms, metastatic abscesses form. These may
occur in any part of the body, but, as a rule, will
follow one of two definite courses. Either they form
in the superficial parts of the body, as in the joints
or subcutaneously; or they occur in the deeper
organs, as in the liver, lungs, and brain. The
formation of each abscess is marked by the occurrence
of rigors. The patient may gradually recover, but
as frequently dies. Death may occur in several
ways:—from exhaustion due to the long-continued
suppuration; from septic pneumonia, peritonitis, or
endocarditis; or from abscesses forming in vital
organs, as the liver and the brain.

Prognosis.—The prognosis is very grave, but it is
not as hopeless as in acute sepsis. The more super-
ficially the abscesses form, the better is the prognosis.
From 50 to 60 per cent. of cases die.

Treatment.—Support the patient's strength in
every way. Administer alcohol in as large quantities
as the patient can be induced to take, and give the
most nutritious and digestible foods. If there be
any local condition of the uterus, as septic endometritis, it should be treated by curetting, and by the
introduction of iodoform. If abscesses form in joints
they should be opened at once, in order to prevent,
if possible, the destruction of the joint. If they form
beneath the skin or muscles, they may be allowed to
point before they are opened. In these cases, as in
acute sepsis, the hypodermic injection of anti-streptococcic serum has been attended by favourable results.

CHAPTER XXX.

PULMONARY EMBOLUS—INVERSIO UTERI—MASTITIS —HÆMATOMA OF THE VULVA—PHLEGMASIA ALBA DOLENS—PUERPERAL INSANITY.

Pulmonary Embolus: Ætiology, Symptoms, Treatment—Inversio Uteri: Ætiology, Symptoms, Treatment—Mastitis: Varieties —Parenchymatous Mastitis: Ætiology, Symptoms—Interstitial Mastitis: Ætiology, Symptoms, Treatment—Hæmatoma of the Vagina or the Vulva: Ætiology, Symptoms, Terminations, Prognosis—Phlegmasia alba dolens: Varieties, Ætiology, Symptoms, Treatment — Puerperal Insanity: Ætiology, Varieties, Symptoms, Treatment.

Pulmonary Embolus.

Pulmonary embolus occurring after delivery is due to the detachment of a clot, most usually from the uterine sinuses, the clot being carried through the right side of the heart into the pulmonary artery.

Ætiology.—Extensive clotting is most likely to occur in cases in which the uterus has not contracted well after delivery. If clotting in the vessels have occurred, any slight movement may be sufficient to determine the detachment of the embolus.

Symptoms.—The onset of the symptoms is extremely rapid. The patient is perfectly well one moment, and the next she is collapsed, asphyxiated, her heart rapid and weak, her breathing frequent and sighing. The duration of the symptoms depends upon the size of the vessel plugged, and upon the strength of the patient. If a large vessel be obliterated, she will die in from a few minutes to a few hours. If the vessel be small, she may recover very gradually.

Treatment.—The patient must be kept absolutely at rest in the recumbent position, and stimulants administered freely. Carbonate of ammonia in large doses is specially recommended as being very rapidly absorbed. Inhalations of oxygen are of doubtful efficacy.

INVERSIO UTERI.

Acute inversion of the uterus is one of the rarest accidents met with in midwifery. The uterus becomes completely, or partly, turned inside out, so that the fundus appears through the cervix.

Ætiology.—The occurrence of inversion is most probable, in the case of a large, lax, thin-walled uterus. It has been caused by:—

(1) Dragging on the placental site by means of the cord.

(2) Violent straining associated with sudden emptying of the uterus;—precipitate labour, severe straining and pressure in the removal of the after-birth (Winckel).

Symptoms.—The occurrence of inversion is followed immediately by the extreme collapse of the patient. There may or may not be severe hæmorrhage.

Treatment.—If the placenta be still adherent, it

should be removed and the uterus immediately replaced by the hand. It must then be thoroughly douched out with creolin solution.

Mastitis.

Mastitis is the term applied to inflammation of the breast.

Varieties.—It occurs in two chief forms :—(1) parenchymatous, (2) interstitial.

(1) **Parenchymatous mastitis.** This is the term applied to inflammation of the milk ducts, *i. e.* of the parenchyma of the breast.

Ætiology.—Parenchymatous mastitis is due to the entrance of bacteria through the nipple. The bacteria may be derived from the milk, which has been allowed to dry upon the nipple, or infection may result from attempts made with septic fingers to form the nipples.

Symptoms.—The first symptom is the appearance of a patch of inflammation over some part of the breast, accompanied by considerable pain. As the inflammation is at first limited to the ducts, and as it is, as a rule, one set of ducts which is infected, so the area of inflammation corresponds in shape to the area from which the affected ducts come. The affected patch is hence triangular in shape, with the apex of the triangle at the nipple, the base at the periphery of the breast. There is a sharp line of demarcation between the healthy and the diseased portion of the breast. The inflammation usually tends to subside, but it may extend into the interstitial substance of the breast.

(2) **Interstitial mastitis**. This is the term applied to inflammation of the interstitial tissues of the breast.

Ætiology.—Interstitial mastitis may start by the extension of a parenchymatous mastitis; or, what is more common, bacteria may find their way directly into the interstitial substance through a crack at the base of the nipple.

Symptoms.—An irregular ill-defined patch of inflammation appears upon the breast; there is intense pain, and severe constitutional disturbance, as, high temperature, rapid pulse, and general malaise. As a rule the affected area suppurates, and an abscess is formed. The presence of pus is recognised not by fluctuation, which is difficult and sometimes impossible to obtain, but by the presence of œdema over the point of suppuration.

Treatment.—The prophylactic treatment of mastitis should be adopted in all cases, but particularly with primiparæ. It consists in hardening the skin of the nipples in order to avoid subsequent laceration (*v.* page 36), and in instructing the patient in the duty of keeping her nipples clean. They should be washed both before, and after, the child takes the breast. Also, milk should not be allowed to accumulate in the breast, if, for any reason, the child stop nursing. This condition, though not in itself sufficient to cause mastitis, still provides a suitable nidus for any germs that may gain admittance. If a crack occur at the base of the nipple, it must be cured as quickly as possible. This is done by touching the crack lightly with nitrate of silver, or better still by painting it twice daily with Tr. Benzoin. Co. If *parenchymatous* mastitis occur, the breast should be firmly

bandaged to the chest-wall, the nipple being first
covered with a small piece of lint soaked in a fifty
per cent. solution of Tr. Benzoin. Co. It is well
also to administer a hydragogue purgative.

If we believe the mastitis to be *interstitial,* and pus
to be likely to form, antiseptic compresses may be
used to prepare the skin for incision. If an abscess
form, it must be opened immediately. The occurrence
of œdema is a positive indication of the presence of
pus ; it is rare to be able to obtain fluctuation. The
following treatment of abscess of the breast is most
successful.—Open into the most dependent part of
the abscess by a radial incision, sufficiently large to
admit the index finger. Let the pus drain out, and
then pass in the finger, and with it break down all
the diseased tissue. By this means the walls of the
loculi in which the pus is stored are broken down, and
one large cavity is formed. Next curette the cavity
with a large curette, choosing one which is not too
sharp, and douche it out thoroughly, so as to wash
away all the débris. Plug the cavity tightly with
iodoform gauze, and bandage the breast as firmly
as possible to the chest-wall.

The gauze must be changed every twenty-four
hours, and the cavity replugged, until the day
comes upon which there is no pus on the gauze.
This date varies from the second to the sixth day
after opening, according to the size of the abscess.
Then the plugging may be discontinued, with the exception of a small piece of gauze, in the skin wound,
in order to keep it open. The breast is bandaged
very tightly, so as to bring the walls of the cavity
into apposition. After this, it need not be dressed for

three or four days. By this time the cavity will have become completely obliterated, and only a small superficial ulcer will remain, which will take a week or so to heal completely. By adopting the above treatment, the worst mammary abscess will be completely healed in from two to three weeks, if care be taken to break down *all* the diseased tissue at the commencement.

HÆMATOMA OF THE VAGINA OR VULVA.

Hæmatoma of the vagina or vulva is the term applied to a collection of blood in the areolar tissues about the vagina or vulva (*v.* Fig. 43).

Ætiology.—As the head descends through the vagina, the return flow of blood through the veins is obstructed, and so the intra-venous pressure is

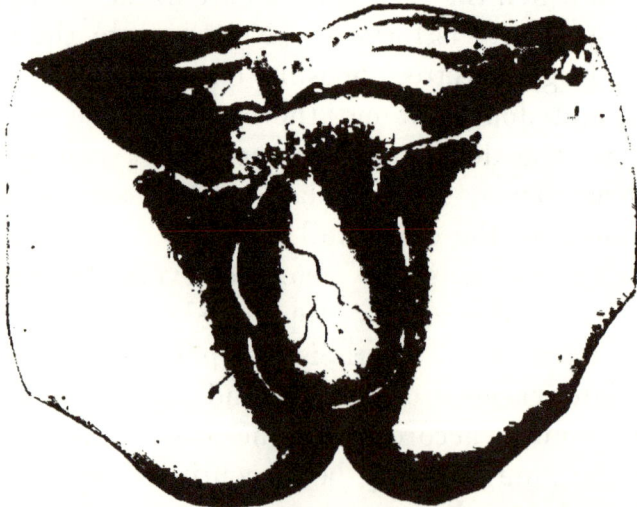

FIG. 43.—Hæmatoma of the left labium. The arrow points to the entrance to the vagina. (Drawn from life.)

increased. Rupture of a vein may then result.

Varicosities of the vulvar or vaginal veins do not appear to predispose to this condition.

Symptoms.—The condition commences with the formation of a small tumour, elastic to the touch and of a blue colour, which gradually increases in size. The vein may rupture before or after delivery, but most usually the condition is not noticed until after delivery. The other symptoms are pain and collapse of the patient, both in proportion to the size of the tumour.

Terminations.—The case may terminate in four ways if the condition remain untreated :—

(1) The tumour may rupture, and free external hæmorrhage result.

(2) The hæmorrhage may extend interstitially upwards towards the abdomen, or downwards towards the perinæum, according as the ruptured vessel is above or below the deep perinæal fascia. The patient may thus " bleed to death into her subcutaneous tissues."

(3) The tumour if small may dry up, and be absorbed aseptically.

(4) Suppuration or decomposition of the contents of the tumour may occur.

Treatment.—If the condition be recognised before delivery, deliver at once. It will usually be possible to apply forceps. If the large size of the tumour obstruct delivery, it must be incised, its contents turned out, and the child delivered past it as rapidly as possible. If the tumour increase in size after delivery, or if it be of large size, it must also be opened, and its contents turned out. In any case in which incision is practised, the cavity should be douched out, and then plugged tightly with iodoform gauze. The plugging is changed every day until the cavity is obliterated. If the tumour be very small, it may be left to absorb. Suppuration should never occur; if it do, the abscess must be opened at the spot at which it points, and free drainage permitted.

Prognosis.—The prognosis depends upon the treatment adopted. The patient may die of hæmorrhage or of sepsis. Neither form of death should occur if the case be correctly treated.

PHLEGMASIA ALBA DOLENS.

Phlegmasia alba dolens, or white leg, is the term applied to the condition which results from thrombosis of one of the veins of the leg.

Varieties.—Two forms occur :— (1) simple or primary, (2) suppurative or secondary.

Ætiology.—The simple form is due to simple thrombosis of the veins of the leg, such as may occur in any condition of debility. Weak action of the heart, and the pressure of an enlarged puerperal uterus on the intra-pelvic veins, combine to favour its occurrence. The suppurative form is secondary to inflammatory changes in the uterine veins. Bacteria travel from the uterine sinuses along the walls of the vein, perhaps by the lymphatic canals. They set up an inflammation in one of the veins of the leg, this inflammation in turn causing the formation of a thrombus.

Symptoms.—The primary form usually comes on a few days after the patient has been allowed out of bed. The leg becomes painful, and the presence of clots in the veins may be determined by running the hand lightly down the back of the calf. If the thrombus form in one of the large veins of the thigh, the limb below the obstruction becomes very much swollen, œdematous, and extremely tender to the slightest touch. In the secondary form pain is felt in one or two distinct foci along the course of the vein. These foci become inflamed and swollen in a day or two, and subsequently a small abscess may form, which points and bursts. The temperature at the same time ranges from 101° F. to 104° F., and the pulse is proportionately rapid.

Treatment.—The treatment of either form consists in absolute rest in the recumbent position. Cold evaporating lotions applied on lint will relieve the pain, in the primary form. The limb should also be wrapped in cotton-wool, and protected by means of a cradle from the pressure of the bedclothes. No friction of any kind must be attempted, for fear of separating a portion of the clot. The secondary form is best treated by hot antiseptic compresses over the inflamed areas.

PUERPERAL INSANITY.

Puerperal insanity is the term applied to a form of madness which sometimes occurs after child-birth. It may last for the remainder of the patient's life, but in the majority of cases is only a temporary affection.

Ætiology.—Insanity may be a primary condition, the result of heredity, alcoholism, or epilepsy ; or it may be merely a symptom of sepsis.

Varieties.—Two varieties occur :

(1) Melancholia.

(2) Mania.

Symptoms.—The symptoms of either variety usually appear in from two to twelve days after delivery. In *melancholia* the patient is extremely depressed, and is frequently found in tears, without any apparent cause. This in itself should be sufficient to direct attention to her condition. If a patient be found to be continually fretting after delivery, without cause, she is probably suffering either from melancholia or sepsis, or perhaps both. She is usually absolutely sleepless and may have various delusions. In *mania* the patient loses all idea of her surroundings, her mind is in a state of chaotic confusion, her moral faculties are affected. One moment she is extremely violent, the next passive and docile. She is the victim of delusions and illusions.

Treatment.—The patient must be watched with the greatest care. She is particularly liable to commit suicide if she be suffering from melancholia. In mania, chloral and bromides may be administered, and purgatives are always required. If any septic condition be present it must be treated.

Prognosis.—More than half the cases recover within six months, but patients are always liable to a relapse during or after subsequent confinements.

CHAPTER XXXI.

INFANTILE FEEDING.

Breast Feeding by Mother and by Wet-nurse—Artificial Feeding —Punctuality—Cleanliness—Suitable Food—Composition of Cow's and of Human Milk—Sterilisation of Milk, Pasteurisation—Farinaceous Foods.

The mother should always suckle her infant herself, unless there be some absolute reason to the contrary. In some cases, either for her own sake, or for the sake of the infant, it may be inadvisable for her to do so. She should not nurse her infant, for her own sake, if she be in a debilitated condition owing to previous hæmorrhages, phthisis, or any other wasting disease. She should not nurse the child, for its sake, if she be suffering from any disease which she may communicate to it, as syphilis or phthisis; also if her milk do not agree with the child, or if her breasts be inflamed. Depressed nipples, or entire absence of nipples, prevent the child from suckling. This may be overcome by improving the nipples in shape, as by drawing them out two or three times a day with clean fingers. If they cannot be improved, a nipple shield may be used, or a *tetarelle*. The latter is an apparatus by the aid of which the mother draws off the milk into a receptacle, from which the child can then suck it. If the mother can be induced to use it intelligently, and if it can be kept clean, it is an excellent instrument. But usually these are insuperable difficulties.

If a mother cannot nurse her child, a wet nurse is the next best substitute. However, it is so extremely difficult to obtain a suitable nurse, that bottle feeding is usually to be preferred. The following are the essentials for a wet nurse :—

(1) She must be perfectly healthy, and free from every disease which might be communicated to the child.

(2) She must be between twenty and thirty-five years of age.

(3) Her breasts must be firm, with well-shaped nipples, and contain abundance of milk.

(4) Her own child must be of the same age as the child she is going to nurse, and must be thriving well upon her milk. Also, she must be prepared to give up nursing it.

(5) Her character must be sufficiently good to allow of her being brought into the patient's house.

Artificial feeding has frequently to be resorted to. If it be properly carried out the child will thrive well, but there are many difficulties in the due performing of it.

There are three essentials in feeding an infant,—punctuality, cleanliness, and suitable food.

Punctuality.—The child must be fed at stated times. It must not be fed between these times, and if it be asleep at the hour for its food it must be awakened. For the first month the child is fed every two hours during the day, leaving one interval of four hours at night. From the commencement of the second month the interval between the meals is gradually increased; until, at the end of the second month, it is being fed every two and a half hours, and at the end of the third month, every three hours.

Cleanliness.—If the child be breast fed, the nipples must be washed with warm water before it is put to them. If the child be bottle fed, the bottle must be kept absolutely clean. A good bottle should be of the familiar boat shape, *i. e.* it should have no angles. The nipple should fit directly on to the mouth of the bottle, without the intervention of a tube. An excellent bottle is now made with an opening at either end, by means of which it can be thoroughly cleaned out. The bottles should be rinsed out with cold water immediately after feeding the child; then scalded with boiling water; and kept, when not in use, in a solution of soda and water. They must be thoroughly rinsed out in cold water before the milk is put into them. Only sufficient milk for the meal should be put into the bottle, and as soon as the child has taken what it wants, the remainder should be poured away.

Suitable Food.—If the child cannot be fed on human milk, then the best substitute is cow's milk in some form. The relative composition of cow's and of human milk is shown in the following table :—

	Cow's Milk.	Human Milk.
Reaction . . .	Faintly acid, due to presence of bacteria.	Alkaline.
Specific gravity . . .	1029	1031
Constituents—		
Fats	3·75	4·13
Lactose	4·42	7·00
Albuminoids . . .	3·76	2·00
(Casein and lactalbumin)		
Inorganic matter . .	0·68	0·20
(Salts, etc.)		
Water	87·39	86·67
	100·00	100·00

19

By referring to this table, we see that cow's milk
differs from human milk in that it contains more
casein, less fat, and considerably less sugar. Not
only does the quantity of albuminoids differ in the
two milks, but also their quality. Cow's milk contains a larger quantity of albumen which is coagul-
able by an acid; consequently, when it is acted
upon by the gastric juice, it tends to form a large
firm curd. Human milk under the same conditions
curdles in a flocculent mass, and so is more easily
digested. We thus see that cow's milk must be considerably modified before it can be made a reliable
substitute for human milk. First of all the proportion of casein must be diminished. This is done by
adding a certain quantity of water, which of course
still further diminishes the proportion of fat and
sugar. These must now be increased by adding sugar
and fat in some form. Demarara sugar is the best
to use, as it counteracts any tendency to constipation
on the part of the child. The amount of fat is
increased by the addition of cream. If cream cannot
be obtained, cod-liver oil may be used instead. Lastly,
we must endeavour to cause the cow's milk to curdle
in a flocculent mass, otherwise this method of feeding
will fail. Barley water, added to the milk instead
of plain water, accomplishes this end. It acts in a
purely mechanical manner, by running between the
drops of milk, which then curdle separately instead of
in one solid mass.

The constitution of a suitable feeding mixture for
a newly-born infant is as follows :—

Milk	.	.	.	One and a half drachms.
Cream	.	.	.	One drachm.
Sugar	.	.	.	Half a teaspoonful.
Barley water	.	.	Five drachms.

The exact amount of barley water to add, and the exact amount of fluid to give the child, can only be found by experiment. On an average, a newly-born infant will take nearly an ounce of the mixture at a time. One part of milk to two parts of barley water is the usual strength to commence with. If the child possit up unchanged milk, it is getting too much fluid. If it pass undigested curds, the milk is too strong. If it digest its food well, but seem always to be hungry, it may get more fluid with proportionately less barley water.

There is still a most important point of difference between artificial food and human milk, *i. e.* that there are always swarms of micro-organisms in cow's milk as we get it. These must be got rid of in some manner. The most obvious method at first sight is to boil the milk. There are, however, great objections to this. It renders the milk less nutritious, and more difficult to digest. Children fed on boiled milk are always constipated. It has been stated, of late, that the nutritive properties of milk are not destroyed unless it be actually boiled; in other words, that any heat short of the boiling point of milk will not change its properties. The method of sterilising milk recommended by Budin, of Paris, is founded upon this theory. It consists in placing the required amount of milk in a bottle which is partially immersed in water. The latter is raised to boiling point, at which it is kept for forty minutes. The bottle of milk is then removed, and rapidly cooled. Dr. Budin says, further, that, if the milk be perfectly sterilised, it does not require to be diluted; in fact, that infants thrive considerably better on undiluted than on diluted milk.

My experience is that infants thrive admirably on a solution of cow's milk prepared as I have described above, and then sterilised for forty minutes in boiling water. I have not found them do so well on the undiluted sterilised milk.

Another method of destroying the germs is by *"Pasteurising"* the milk. This consists in raising it to a temperature between 158° F. and 176° F., and keeping it at this temperature for thirty to forty minutes. This method will destroy the greater number of bacteria and spores, but not all.

Farinaceous foods containing starch should never be given to infants, as the secretions by which starch is digested, *i. e.* the saliva and pancreatic juice, are not fully established until the child is six months old. Condensed milk, and prepared foods in which the starch has been changed into sugar, are also for the most part objectionable; they almost all contain too much carbohydrate, and too little nitrogenous matter. As a result the infants become large and fat, but have not sufficient development of bone and muscle. They may, if necessary, be used for the first two months of the child's life, but after that more nitrogenous food must be substituted.

CHAPTER XXXII.

SOME INFANTILE DISEASES.

Green Diarrhœa: Ætiology, Symptoms, Treatment—Thrush: Ætiology, Symptoms, Treatment—Ophthalmia Neonatorum: Ætiology, Symptoms, Treatment—Asphyxia Neonatorum: Varieties, Symptoms, Treatment, Schultze's Method of Artificial Respiration — Icterus Neonatorum: Ætiology, Treatment—Late Hæmorrhage from the Cord: Treatment— Cephalhæmatoma.

GREEN DIARRHŒA.

DIARRHŒA, in the infant, consists in the passage of more than six stools in the twenty-four hours. The normal motions of an infant are yellow in colour, liquid in consistency, and slightly fæcal in odour.

Ætiology.—Green diarrhœa is directly due to the entrance of bacteria into the child's stomach. They gain access in the food, most usually in sour milk. The green colour of the stools is due to a pigment which is formed by them in the stomach.

Symptoms.—The diagnostic symptom is the passage of green motions, and masses of foul-smelling, semi-digested curds. If the case be allowed to remain untreated, gastritis results, which may extend downwards into the intestines. The child then becomes marasmic, as it is unable to assimilate its food, and dies of starvation.

Treatment.—The prophylactic treatment consists in giving the child milk which is free from germs, and in keeping its bottle perfectly clean. If green diarrhœa occur, the indication is to clear the curds out of its stomach. The administration of a teaspoonful of castor oil will usually be found to be sufficient, if the case be taken in time. Sometimes, however, this will not suffice, and then more radical measures must be taken. Stop all milk, and feed the child with raw beef juice and barley water. By this means the bacteria are starved out, so to speak, as the majority of them can exist only on milk. This line of treatment, accompanied by the administration of castor oil, is continued for two or three days. Then, if the diarrhœa have ceased, the child may return to its ordinary diet. If the child be very weak and marasmic, it requires stimulants. This is best given in the form of white wine whey.

THRUSH OR STOMATITIS MYCOSA.

This is another disease which results from sour milk. It is the term applied to the formation of small white spots on the mucous membrane of the mouth and tongue.

Ætiology.—It is directly due to the implantation of a fungus, *Oidium albicans* (by some *Saccharomyces albicans*), on the mucous membrane of the mouth. The Oidium is found in sour milk; and, thus, the child may become infected from milk which has decomposed upon the mother's nipple, or in a dirty bottle.

Symptoms.—Small white spots appear on the mucous membrane of the mouth. If untreated, the spots coalesce and form a species of false membrane, which may extend into the pharynx and œsophagus. Green diarrhœa is frequently associated with this condition.

Treatment.—The prophylactic treatment consists in washing the mother's nipples before the child takes the breast ; in having the bottle perfectly clean, if the child be bottle-fed ; and in carefully wiping the mouth of the child with a soft rag, after it has had its food. If thrush occur, treat it at once. Place a teaspoonful of glycerine of borax (B.P.) in the child's mouth, three times a day. This acts as an antiseptic, and destroys the fungus.

OPHTHALMIA NEONATORUM.

Ophthalmia neonatorum is an infectious disease of the eyes, with which the child may become inoculated during the passage of the head through the vagina.

Ætiology.—It is almost always due to the entrance of Neisser's gonococcus into the eyes, most usually during the passage of the head through the vagina. It has, however, occasionally been found to have resulted from the entrance of other forms of bacteria.

Symptoms.—Two days after birth, *i. e.* after infection, the symptoms commence. The eyelids become swollen and inflamed, and a purulent discharge flows from between them. In severe cases, opacities or even ulcers of the cornea may form, and so partial or complete loss of vision result.

Treatment.—Prophylactic treatment should be adopted as a routine in hospitals, also, occasionally, in private practice, wherever there is any reason to suspect gonorrhœal infection in the mother. It consists in wiping the eyes of the child carefully with a soft rag the moment the head is born, and then in dropping in a solution of nitrate of silver of a strength of four grains of the nitrate to an ounce of water.

If the infection occur, our treatment must be more active. The eyes must be well washed with warm water five or six times a day, the lids being separated so as to allow the pus to flow out. At the same time a two per cent. solution of nitrate of silver is dropped into the eyes once a day; and they must be kept carefully bandaged. The greatest care must be taken to avoid spreading the infection by means of dirty fingers or cloths. If only one eye be infected, the sound eye must be treated with the weaker solution of nitrate of silver. It must also be protected from infection by a hermetic bandage. To do this, apply a small piece of lint spread with boracic ointment to the eye, then a pad of cotton wool, and cover the whole with strips of lint soaked in collodion (Swanzy).

ASPHYXIA NEONATORUM.

Infants are frequently born asphyxiated after protracted labour, or when a malpresentation occurs, especially if it be a breech.

Varieties.—There are two distinct varieties :—

(1) Asphyxia pallida, or white asphyxia.
(2) Asphyxia livida, or blue asphyxia.

The worse form of asphyxia is asphyxia pallida. In it the child is perfectly white when born; the cord is not pulsating; the heart can be barely felt; there are no attempts at respiration; and all reflexes are lost.   In asphyxia livida the child is blue; the cord pulsates; the heart beats strongly; there are slight attempts at respiration; and the reflexes are present.   In order to feel an infant's heart, press the fingers up under the arch of the ribs, a little to the left of the sternum.   It is then easily felt, if it be beating.

Treatment.—To treat an asyphyxiated infant successfully, a regular line of action must be laid down and carried out in due order, paying the greatest attention to details.   If the child be born in white asphyxia,—

(1) Ligature and divide the cord.

(2) Place the child in a bath of water at 100° F.

(3) While it is in the bath, suck the mucus out of the trachea with a silver or gum-elastic catheter.

(4) Take the child out of the bath, and dry it thoroughly.

(5) Perform Schultze's method of artificial respiration (see below) five or six times.

Repeat steps (2) to (5) over and over again until either the child dies, *i. e.* its heart stops, or until it passes into the stage of asphyxia livida.   As soon as this occur, we may assume that its reflexes have returned, and may endeavour to stimulate them.   To do this, after taking the child out of the hot bath, plunge it for a moment into a cold bath; then " Schultze " as before.   Continue this routine,—hot bath, extraction of mucus, cold bath, dry, " Schultze,"

until the child commences to make strong efforts at
respiration.   Then if there be a fire in the room, make
the nurse sit down in front of it, and roll the child
on her knees.   To do this she places the child across
her knees on its side, and then rolls it half over on to

FIG. 44.—Schultze's method of performing artificial respiration.
Completion of the inspiratory movement.

its chest, at the same time compressing the ribs.   This
causes expiration.   She then rolls it in the opposite
direction on to its back ; at the same time removing
all pressure from the chest, and pulling upon the
arm which is uppermost, in such a way as to draw

the ribs upwards. This causes inspiration. It is also a good thing to rub whisky on the gums and chest of the child. A child in white asphyxia must not be placed in a cold bath, as it depresses the heart dangerously.

Schultze's method of artificial respiration is performed as follows : *—Seize the child in both hands, the thumbs hooked beneath the heads of the humeri, the index fingers along the sides of the thorax, the

FIG. 45.—Schultze's method of performing artificial respiration. Expiration.

other three fingers along the back (v. Fig. 44). Then raise the child with a quick sweep through the air until its body rolls forward upon your thumbs, which are now placed on the anterior aspect of the chest

* This description differs slightly from the original description of Schultze. He recommends to hook the index fingers, and not the thumbs, under the axillæ of the child, thus avoiding the necessity of changing the grip.

(*v.* Fig. 45) ; and at the same time compress the chest laterally with the index fingers and posteriorly with the other fingers, so diminishing its lateral and antero-posterior diameters. Owing to the position of the child, the abdominal viscera fall towards the diaphragm, forcing it upwards, and in this way diminish the vertical diameter of the chest. This movement causes expiration, and the inverted position favours the flow of mucus out of the trachea. Having kept the child for a moment in this position, it is then swung forwards again into a vertical position. As the child falls forward all compression is removed from the chest and the child held by the shoulders ; so that, as it falls, its weight causes the ribs to be drawn upwards (*v.* Fig. 44). This movement causes inspiration.

If the infant be born in a state of asphyxia livida, the cord must not be tied until it has ceased pulsating. It is then tied, and the child treated as described.

ICTERUS NEONATORUM.

Icterus neonatorum, or infantile jaundice, occurs as a symptom in three conditions :—

(1) Physiological or simple icterus occurs in a large number of infants. It is said to be due to the destruction of large numbers of red blood-corpuscles, after birth, which causes an increase in the amount of bile formed in the liver.

(2) Severe or malignant icterus occurs in cases of inflammation of Glisson's capsule ; the latter being usually due to extension of inflammation from the umbilicus.

(3) Icterus also occurs due to congenital diseases, as syphilis or malformation of the liver.

Treatment.—In simple icterus the child requires no special treatment. A mild laxative, such as glycerine, may be given in order to clear out the digestive tract. In malignant icterus the prognosis is very bad. If there be any septic condition of the umbilicus, it must be treated with antiseptic compresses or iodoform powder. The child should also be given stimulants.

LATE HÆMORRHAGE FROM THE CORD.

Late or secondary hæmorrhage from the cord may occur any time during the fortnight subsequent to delivery.

Ætiology.—It may be due to syphilis, hæmophilia, acute fatty degeneration, hæmoglobinuria (Winckel) ; but the commonest cause is ulceration of the umbilicus due to septic infection.

Treatment.—It is extremely difficult to check, and in most cases fatal. Perchloride of iron, ligature of the entire umbilical ring, plugging of the umbilical fossa with plaster of Paris, and pressure of all kinds, have been tried without avail. The method most recommended consists in attempting to underpin the umbilical vessels with a stout needle, and then compressing them against the needle, by passing a figure-of-eight ligature beneath its projecting ends.

CEPHALHÆMATOMA.

Cephalhæmatoma is the term applied to an extravasation of blood, which sometimes forms, during labour, under the periosteum subjacent to the caput succedaneum. It is due to the rupture of a vessel during delivery. At first it consists of a tense, slightly fluctuating tumour, and is limited in extent by the sutures which surround the bone over which it forms. As the blood coagulates, the periphery of the swelling becomes as hard as bone, while in the centre there is a depression. At this stage it feels exactly as if there were an opening through one of the bones into the skull. It should not be interfered with unless it suppurate, and then it must be opened and drained.

APPENDIX.

———

THE first of the following tables shows the nature and the proportion of the cases treated in the Rotunda Lying-in Hospital, during the mastership of Dr. W. J. Smyly. The second table shows the number of deaths that have occurred during the same period, and their cause.

TABLE A.—*Showing the Nature of the Cases treated in the Rotunda Lying-in Hospital.*

	1889-90.	1890-91.	1891-92.	1892-93.	1893-94.	1894-95.	1895-96.	Total.	Average.
Total number of labours	1199	1184	1219	1288	1316	1267	1524	8997	
Abortions	29	28	50	41	50	40	44	282	1 in 31.9
Placenta prævia	6	5	6	8	7	14	9	55	1 in 163.58
Accidental hæmorrhage	11	13	20	12	11	5	2	74	1 in 121.58
Post-partum hæmorrhage	23	14	11	24	14	18	16	120	1 in 74.97
Secondary post-partum hæmorrhage	3	5	—	1	—	—	—	9	1 in 999.6
Hæmatoma of vulva	—	—	—	—	—	—	2	2	1 in 4498.5
Hyperemesis	—	2	1	1	1	1	—	6	1 in 1499.5
Hydramnios	6	6	5	3	2	3	1	26	1 in 346.03
Myxoma chorii	—	—	4	—	1	1	—	6	1 in 1499.5
Eclampsia	5	6	6	9	4	—	3	33	1 in 272.63
Insanity { Mania	3	—	2	3	—	—	3	11	1 in 817.9
Insanity { Melancholia	—	—	1	1	—	—	—	2	1 in 4498.5
Pelvic presentations	41	48	29	43	23	49	48	281	1 in 32.01
Face	3	4	3	3	4	6	—	23	1 in 391.17
Cross-births	3	1	2	3	4	3	4	20	1 in 449.85
Brow	1	1	2	5	1	2	3	15	1 in 599.8

	1 in 63·1	1 in 131·5	1 in 152·375	1 in 75·76	1 in 188	1 in 253·4	1 in 1524	Total
Prolapse of cord	2	8	14	12	6	7	7	56 — 1 in 160·66
Induction of premature labour	2	2	3	6	3	7	3	26 — 1 in 346·03
Adherent placenta	15	10	10	19	10	10	17	91 — 1 in 98·86
Version	8	11	6	11	14	16	13	79 — 1 in 113·88
Forceps	47	22	38	45	41	32	42	267 — 1 in 33·69
Craniotomy	3	3	2	1	1	1	2	12 — 1 in 749·75
Decapitation	—	—	—	—	1	—	—	1 — 1 in 8997
Cæsarean section	—	2	—	—	1	—	—	3 — 1 in 2999
Porro's operation	—	1	+	—	1	—	1	3 — 1 in 2999
Symphysiotomy	—	—	1	3	—	1	1	4 — 1 in 2249·25
Rupture of uterus	—	—	—	—	—	—	—	2 — 1 in 4498·5
Rupture of cervix and vagina involving Douglas's pouch	—	2	1	—	7	5	—	3 — 1 in 2999
Maternal mortality	19	9	8	17	7	5	1	66 — 1 in 136·31
Average maternal mortality	1 in 63·1	1 in 131·5	1 in 152·375	1 in 75·76	1 in 188	1 in 253·4	1 in 1524	1 in 136·31
Percentage maternal mortality	1·58	0·76	0·65	1·32	0·53	0·39	0·06	0·73

20

TABLE B.—*Showing Cause of Death in the Rotunda Lying-in Hospital.*
1889–90. TOTAL 19.

Name.	Admitted.	Delivered.	Died.	Cause of death.	Notes.
C. R.	Nov. 30	Nov. 30	Dec. 1	Eclampsia	Chloral and chloroform treatment.
S. C.	Dec. 8	Dec. 8	Dec. 8	Accidental hæmorrhage	Child perforated and extracted.
M. C.	Dec. 19	Dec. 19	Dec. 19	Accidental hæmorrhage	Accouchement forcé.
T. H.	Dec. 28	Dec. 28	Dec. 28	Phthisis	Admitted moribund.
T. K.	Dec. 27	Jan. 5	Jan. 10	Septicæmia	Induction of premature labour by Barnes' bags.
J. S.	Jan. 14	Jan. 14	Jan. 15	Eclampsia	Chloral and chloroform treatment and delivery by craniotomy.
S. H.	April 8	April 13	April 14	Purulent meningitis	Suffered from influenza previous to admission. Autopsy.
M. B.	April 28	April 30	May 1	Eclampsia	Chloral and chloroform treatment.
L. F.	May 6	May 7	May 23	Pyæmia	No operation performed.
A. F.	June 10	June 11	June 22	Septicæmia	No operation performed.
E. C.	June 20	June 22	June 25	Phthisis	Admitted in last stage.
A. B.	June 18	June 19	July 25	Septicæmia (?)	No operation performed.
E. M.	Aug. 2	Aug. 2	Aug. 27	Pyæmia	No operation performed.
M. B.	Aug. 15	Aug. 17	Aug. 29	Septicæmia	No operation performed.
S. P.	Aug. 16	Aug. 23	Sept. 5	Septicæmia	P.M.—Large sloughing myoma uteri.
M. L.	Aug. 18	Aug. 20	Sept. 25	Pyæmia	No operation performed.
M. F.	Oct. 6	Oct. 7	Oct. 12	Septicæmia	Delivery by forceps.
B. M.	Oct. 15	Oct. 15	Dec. 22	Pyæmia	No operation performed.
C. O'N.	Oct. 24	Oct. 29	Oct. 29	Intestinal obstruction	Suffering from intestinal obstruction for some days previous to admission.

1890–91. TOTAL 9.

M. McG.	Jan. 4	Jan.	Jan. 9	Phthisis	Admitted in last stage.
M. R.	Jan. 10	Jan. 10	Jan. 13	Pneumonia	Admitted with croupous pneumonia.
J. L.	Feb. 1	Feb. 1	Feb. 1	Accidental hæmorrhage	Accouchement forcé.
C. M.	Mar. 2	Mar. 2	Mar. 2	Accidental hæmorrhage	Accouchement forcé.
E. C.	May 5	May 5	May 6	Epilepsy	History of epilepsy for years.
M. F.	Aug. 4	Aug. 5	Aug. 6	Hyperemesis	Vomiting for a month previous to admission.
B. B.	Aug. 5	Aug. 6	Aug. 9	Pneumonia	Admitted with symptoms of pneumonia.
M. B.	Aug. 6	Aug. 6	Aug. 6	Hæmorrhage, rupture of uterus	Marginal placenta prævia; forceps delivery; died on couch.
C. K.	Oct. 2	Oct. 3	Oct. 3	Rupture of uterus	Long-standing case of prolapse of cervix; uterus ruptured 6 hours after labour commenced.

1891–2. TOTAL 9.

A. P.	Nov. 3	Nov. 4	Nov. 4	Eclampsia	Chloral and chloroform treatment.
M. S.	Nov. 18	Nov. 20	Nov. 29	Mania	Unmarried. Temperature remained normal until 30 minutes before death, then rose to 109·4° F.
B. C.	April 8	April 9	April 9	Mitral disease	Admitted in last stage.
M. M.	April 14	April 14	April 15	Empyema	Chloral and chloroform treatment.
C. A.	June 14	June 14	June 15	Eclampsia	Admitted with croupous pneumonia.
A. C.	June 23	June 24	June 26	Double pneumonia	Admitted with rupture; Porro's operation performed, but it failed to check hæmorrhage.
M. R.	June 29	June 30	June 30	Rupture of cervix and vagina	
M. B.	July 6	July 6	July 7	Pneumonia	Admitted with broncho-pneumonia.
M. B.	Oct. 27	Oct. 27	Oct. 28	Eclampsia	Chloral and chloroform treatment.

1892–3. Total 17.

Name.	Admitted.	Delivered.	Died.	Cause of death.	Notes.
M.	Dec. 17	Dec. 18	Dec. 18	Eclampsia	Chloral and chloroform treatment.
M. M'C.	Dec. 17	Dec. 18	Dec. 18	Ruptured cervix	Admitted with ruptured cervix owing to unskilled use of forceps outside. Died of hæmorrhage.
M. N.	Jan. 1	Jan. 27	Jan. 27	Eclampsia	Chloral and chloroform treatment.
M. A. H.	Mar. 9	Mar. 10	Mar. 10	Phthisis	Admitted in last stage.
M. M. D.	April 7	April 7	April 8	Septicæmia	Admitted in advanced stage of sepsis.
M. F.	April 2	April 3	April 21	Pulmonary embolism	Occurred 18 days after delivery; subsequent to phlebo-thrombosis in leg.
M. H.	May 31	May 31	June 1	Post-partum hæmorrhage	Myomatous uterus.
M. P.	June 19	June 19	June 19	Mitral stenosis	Admitted in last stage.
M. M. C.	June 24	June 28	July 1	Cerebro-spinal meningitis	Ill before admission.
L. C.	June 25	June 25	July 14	Pyæmia	Ruptured symphysis during labour, and an abscess formed between the bones. Uterus normal.
C. B.	July 18	July 18	July 18	Accidental hæmorrhage	Vagina plugged until labour set in, then accouchement forcé.
M. O'C.	July 24	July 24	July 29	Sepsis	Admitted with ruptured cervix due to improper use of forceps; also septic.
M. A. R.	Aug. 31	Undelivered	Sept. 9	Uræmia	Admitted septic.
K. L.	Sept. 24	Sept. 24	Sept. 26	Peritonitis	Admitted septic.
S. W.	Sept. 25	Sept. 25	Oct. 20	Mania	
K. D.	Sept. 30	Sept. 30	Sept. 30	Accidental hæmorrhage	Admitted with the vagina plugged, though membranes were ruptured; perforation followed by extraction; died on couch.
L. M'G.	Oct. 29	Oct. 29	Nov. 9	Sapræmia	Child delivered by symphysiotomy; wound sloughed.

1893–4. TOTAL 7.

M. D.	Nov. 13		Nov. 20	Uræmia	Child putrid; physometra.
K. C.	Dec. 20	Dec. 29	Jan. 5	Septicæmia	Morphia treatment.
E. D.	Feb. 19	Feb. 19	Feb. 24	Eclampsia (?)	Patient delivered by forceps; no hæmorrhage; died suddenly. P.M.—All organs healthy; no laceration of uterus or vagina.
M. K.	Aug. 18	Aug. 18	Aug. 19	Syncope	
K. K.	Aug. 15	Aug. 16	Aug. 19	Hyperemesis	Died of exhaustion.
E. H.	Oct. 1	Unde-livered	Oct. 6	Septicæmia	Plugged outside for accidental hæmorrhage.
J. S.	Oct. 28	Oct. 28	Oct. 30	Cæsarean section	Died of ruptured intestine, which was injured during operation.

1894–5. TOTAL 5.

M. H.	Dec. 23	Dec. 23	Dec. 25	Eclampsia	Morphia treatment.
E. L.				Particulars lost	
C. H.	June 25	July 1	July 2	Bright's disease	Hyperemesis for 6 weeks before admission.
E. D.	Aug. 31	Unde-livered	Sept. 9	Heart disease	Died undelivered and not in labour.
M. B.	Oct. 11	Unde-livered	Oct. 13	Bright's disease	Died undelivered and not in labour.

1895–6. TOTAL 1.

M. D.	Feb. 7	Feb. 16	Feb. 25	Septicæmia	Labour induced by bougies.

Recipes for Infant Feeding.

Beef Juice.

To prepare beef juice, take four ounces of lean beef, cut it up into small pieces, and place it in a jar with four ounces of cold water. Let the whole stand in a cool place for two and a half hours, then strain the mixture. Give the infant one part of the juice, thus obtained, diluted with two parts of barley water. To avoid decomposition and the formation of ptomaines, the beef juice must be kept in a cool place, and prepared fresh for each meal.

Barley Water.

Add two teaspoonfuls of well-washed pearl barley to a pint of cold water. Boil the mixture until it is reduced to three quarters of a pint.

Humanised Milk.

Pure new milk	.	One and a half drachms.
Cream	.	One drachm.
Demerara sugar	.	Half a teaspoonful.
Barley water	.	Four and a half drachms.

Pour the mixture into a bottle, the neck of which is plugged with cotton wool. Place it in a saucepan of warm water over a fire or gas stove, and allow it to remain there for forty minutes after the water has commenced to boil. Then remove the bottle, and cool rapidly.

White Wine Whey.

Add two ounces of sherry to half a pint of pure new milk, which has been heated to boiling-point. Bring the milk again to the boil for a moment. The mixture curdles, and the whey is separated by straining through muslin. The infant may be given up to half an ounce of the whey at a time.

Lime Water.

Add a lump of unslaked lime to a pint of water. Mix it thoroughly, and then allow the mixture to stand for a couple of hours. As soon as the excess of lime has fallen to the bottom of the vessel, pour, or syphon off, the clear fluid.

INDEX

PRINTED BY ADLARD AND SON,
BARTHOLOMEW CLOSE, E.C., AND 20, HANOVER SQUARE, W.